I, WARBOT

KENNETH PAYNE

I, Warbot

The Dawn of Artificially Intelligent Conflict

OXFORD
UNIVERSITY PRESS

OXFORD
UNIVERSITY PRESS

Oxford University Press is a department of the
University of Oxford. It furthers the University's objective
of excellence in research, scholarship, and education
by publishing worldwide.

Oxford New York
Auckland Cape Town Dar es Salaam Hong Kong Karachi
Kuala Lumpur Madrid Melbourne Mexico City Nairobi
New Delhi Shanghai Taipei Toronto

With offices in
Argentina Austria Brazil Chile Czech Republic France Greece
Guatemala Hungary Italy Japan Poland Portugal Singapore
South Korea Switzerland Thailand Turkey Ukraine Vietnam

Published in the United States of America by
Oxford University Press
198 Madison Avenue, New York, NY 10016

Library of Congress Cataloging-in-Publication Data is available
Kenneth Payne.
I, Warbot: The Dawn of Artificially Intelligent Conflict.
ISBN: 9780197672358

Printed in the United Kingdom on acid-free paper
by Bell and Bain Ltd, Glasgow

For Stephen Hare and Stephanie Jones

'There is a play of possibilities, probabilities, good and bad luck, which makes war of all branches of human activity the most like a game of cards.'

Carl von Clausewitz, *On War*, 1832

'*We tried everything we could and it was just too strong.*'

Elite poker player Jason Les, defeated by Libratus AI, 2017

CONTENTS

INTRODUCTION

A CALL TO ARMS

AI is going to war. From the chaos of battle to the quiet hum of headquarters far from the action, intelligent machines are about to change conflict forever. Rapid advances in Artificial Intelligence are disrupting manufacturing, healthcare, transport, finance, retail and more. Now they've begun to transform the equipment and techniques of modern warfare. Whether it's combat droids dog-fighting far overhead or AI advising politicians and generals, we are on the cusp of artificially intelligent warfare. Russia's Vladimir Putin reckons that 'whoever leads in AI will rule the world'.[1] It's just possible, I suppose, that he's not thinking about war.

Weapons that can think for themselves and respond to changes in their environment are hugely valuable. They spare the lives of our own soldiers and they will soon deliver better performance than crewed systems—shooting accurately and manoeuvring swiftly. Perhaps that's why there's always been a close connection between AI research and the military, one that continues today. It's not just about weapons. AI is a decision-making technology, which certainly makes it useful for weapons—for uninhabited tanks, aircraft and ships, and guided missiles—but also for strategy, the thinking part of warfare, where commanders weigh what to do. In battle we want warbots that will outperform the enemy. Speed and accuracy of decision-making are imperative, and AI is blisteringly quick. Much faster certainly than humans: already an algorithm has repeatedly outclassed an experienced fighter pilot in simulated aerial combat.

Strategy, however, calls for different qualities. Not for nothing are countless books called 'the art of war'. Creativity, imagination, guile—all are valued attributes. Strategy requires good judgment, even though there are innumerable variables interacting simultaneously. Astute strategists strive to understand what the enemy is thinking. Should we escalate? How much force will it take to deter an attacker? These are harder, less tightly bounded decisions than those facing the AI fighter pilot, or even the networked system that controls a swarm of them. Here too, though, modern AI has an important part to play in conflict, and it's one that will only grow in the near future. All of which is making many observers uneasy. Can we trust machines to wage war? Can we afford not to? What chance is there of regulating 'warbots', intelligent warfighting machines, especially in view of Putin's disturbing comment?

If it's boom time for AI, it's also boom time for writing about AI, including in war. And yet there have been vanishingly few attempts to seriously engage with the military impact of Artificial Intelligence. Much attention in discussions of warbots focuses on capabilities of individual platforms and technologies—the drones or missiles, the high-definition cameras and algorithms that monitor them—rather than the ways they'll be employed together and their deeper impact on warfare. Instead, most public discussion of military AI majors on whether autonomous weapons should be banned altogether. While that is a serious question, which we'll consider too, it often floats free of a larger discussion about why states want to acquire such weapons in the first place, and how they intend to employ them.

INTRODUCTION

The central argument of this book is new thinking is required for a new era of AI weapons. In itself, that might not be a novel argument; indeed you can read similar claims in many thinktank reports and newspaper op-eds. But the analysis typically stops there, caught up in discussion of unethical 'killer robots' or the latest privacy-invading facial recognition algorithm.

What's sorely needed is a deeper understanding of the military possibilities that modern AI allows, and also its limitations. These in turn should be firmly grounded in the cultural history of AI development. Technology isn't just 'out there', somehow emerging from the ether to transform warfare, but is imagined and then created by actors whose perspectives are shaped by their experiences. And there's another aspect to consider carefully: the psychological essence of military activity. AI is a decision-making technology, rather than a weapon. To my mind, that's precisely what makes it such a radical proposition—for the entirety of our existence as a species, we humans have decided the issues about which we fight, and the ways in which we go about it. Tactics in battle and strategy ahead of it are both intimately connected to our biological minds. That is no longer entirely true—now decisions about using force are being made by other means. Grasping the distinction between human and AI decision-making is essential for understanding how conflict will be shaped by the advent of sophisticated warbots.

And so, in successive chapters, we'll first explore the ways in which Artificial Intelligence and war have acted on one another, from the very beginning at Bletchley Park in the Second World War through to the present day. Then we'll look ahead to the battlefield of the near future. It's an environment where today's AI will be extremely useful—its relentlessness, accuracy and speed of decision-making are invaluable assets. AI will prompt the invention of new equipment, and the development of new concepts through which these might be used. Why have a big, heavily armoured tank when there's no crew to protect and when you can target the enemy with missiles fired with pinpoint accuracy from hundreds of miles away? Why risk a huge, lumbering aircraft carrier, just to project a handful of strike aircraft against enemy targets, when a drone swarm can be launched from an crewless submarine patrolling offshore?

Next we'll turn to the bigger picture—the way in which the action is coordinated at the strategic level. AI might now be brilliant at chess, but no board game is anything like as complex as military strategy. Still, there are reasons to suspect that in the near future AI will have a significant role in decision-making at the strategic level too. And then we'll return to the issues of control, ethics and regulation—to the challenge of making sure that our warbots serve us faithfully, and that we in turn deploy them responsibly. Equipped with a deeper understanding of where warfighting AI comes from and what it offers, we should be better placed to address the issues that rightly concern those campaigning to ban warbots.

How intelligent is AI?

The Artificial Intelligence of today would astonish a visitor from the recent past. The science fiction writer Arthur C. Clarke wrote in the 1970s that any sufficiently advanced technology is indistinguishable from magic. That might well be the experience of someone catapulted forward in time half a century from then to today. Here we are, on the cusp of self-driving cars, where the smartphone in everyone's pocket responds to voice commands by pulling up useful information from the ether. That's precisely the sort of magic that Douglas Adams imagined in his bestselling 1970s novel *The Hitchhiker's Guide to the Galaxy*. It's now almost a decade since an AI called Watson thrashed humans at *Jeopardy*, the popular television general knowledge quiz. AIs can outperform hugely experienced clinicians with decades of training in searching for tumours on x-ray images. And then there's war. On the decks of American aircraft carriers patrolling the Pacific Ocean, experimental AI aircraft warm their engines. And under the waves, Artificial Intelligence is powering the next generation of submarines.

The flip-side of magic and wonderment is banality. As with most technological development, we quickly accommodate ourselves to the latest thing, which rapidly becomes normal and even mundane. Is today's AI actually intelligent? Once we've assimilated what it does, the magic dissipates. Perhaps this isn't proper intelligence after all, but just a souped-up calculator. Is the navigation computer in your family saloon really AI? Then why has it sent you down this tiny, winding one-way lane? And anyway, while it might be good at one task, today's AI is really only good at that: it's not as though your car's satellite navigation can beat you at chess, still less try its hand at interior design. Yes, AI might perform impressively at that one particular skill, but it's still worse than a toddler at the sort of tasks we humans take for granted—specifically, it lacks common sense, and the ability to generalise from one task to another. Scholars say that modern AI is both narrow and brittle—good at one (narrow) thing, but easily reduced to producing nonsense when conditions change even slightly from optimal (brittle). That might be a profound drawback in war, where conditions are chaotic and uncertain, and where the unexpected is apt to happen. Our AI is lightning quick and has unfailing memory, especially for numbers. But it's often dumb as a brick—just ask Siri, Apple's voice assistant. You'll probably get nonsense by way of reply, or, just as telling, a witty response that's clearly been handcrafted by a human.

Creativity and imagination are the hallmarks of human intelligence, and vital in war too. AI might be quick on the draw, a useful trait for autonomous weapons, but it can't think like a brilliant artist, or even an average one. When was the last great AI novel you read? Are you listening to AI music? It's unlikely, because even state-of-the-art AI is not good at these things. It lacks an essential human quality. Scholars sometimes use the term 'uncanny valley' to describe AI's creative efforts—they can be almost convincing, but there's still something odd about them that produces an uncomfortable sensation.

Which is the right view of AI? Superhuman brain, or outsized calculator? Reality lies somewhere in the middle. Today's AI is capable of truly astonishing feats. It can produce realistic translations of foreign language—spoken and in text. It can recognise what's in a photograph—a cat, a frisbee, a beach. But on the other hand, it is still a long way short of human intelligence, with its signature ability to flexibly adapt to new circumstances. Homo sapiens, the one who knows, is still the ultimate intelligence.

Science fiction often anticipates an imaginary future where incredibly intelligent machines and humans interact—often to the detriment of humans. But for now that's an impossibly distant goal. Some AI visionaries anticipate a 'singularity'—the emergence of a superintelligence far surpassing our own. They debate the 'existential risk' to humanity of this sort of AI. Elon Musk, the guru of SpaceX and Tesla fears that we are 'summoning a demon'.[2] The late Stephen Hawking warned that AI could 'spell the end of the human race'.[3] It's a disturbing vision—but the visionaries themselves can't agree when it might happen. Those troubled by the prospect include some of the world's leading AI researchers. A 'super-intelligent' computer, far smarter than humans, and with desires that run counter to ours might spell doom for humanity. A malign superintelligence wouldn't need eight billion humans, competing for scarce resources, though it might munificently keep some of us on as zoological specimens. Lacking emotion and empathy, why would it want to do otherwise? After all, that array of biomass could be put to a more efficient use. The superintelligence might even think it a bit rich for humans to complain, since we've already devastated our own environment. As we unpick the differences between AI and human psychology, I hope you'll gain some perspective on just how unlikely this threat is.

A far more pressing danger exists—one that doesn't need science fiction breakthroughs. The technology required is already here, or at least not far distant. A more realistic prospect than super-intelligent AI overlords is a machine with lethal powers, considerable autonomy and much more limited intelligence. This is entirely plausible—as we'll see, it's all but inevitable. Such machines exist already and have been used in combat. Many more will soon join them, and collectively they'll control ever-greater lethal force. This raises many troubling ethical issues. Will a machine be able to distinguish between enemy combatants (legitimate targets during wartime) and civilians? That might be particularly tricky in irregular conflict, especially where guerrilla forces deliberately seek to hide among the general population. Will the machine use the appropriate level of violence—avoiding unnecessary escalation, which may in turn prompt retaliation, and so beget an automated spiral of violence? Who will be in command if a system is fully autonomous? Can they call it back once it's underway? What if they change their mind about the instructions they issued? And who is ultimately responsible if it does the wrong thing? Also, some ask, does it not strip humans of their dignity to be killed by emotionless robots? War is always terrible, but its essence is human. Is it not deeply unethical to exterminate people, perhaps from many miles away, without any human being involved in the decision?

I too am concerned about the ethical implications of warbots although I'm sceptical that a way will be found to ban them. They are simply too useful. States fear being left behind if they don't develop effective AI. Even if they did manage to settle on a definition of a warbot (tricky, as we'll soon see), it would be hard to enforce any ban. With nuclear weapons, there's often a trail of suspicious activity to follow. A clandestine nuclear programme, like North Korea's, makes plenty of noise. There are rocket tests, and underground explosions; there are industrial sites where the ingredients are enriched; and there is global espionage as nuclear wannabes seek out blueprints, experts for hire and raw ingredients. None of that holds for AI weapons, at least to the same extent.

So, we ought to be thinking very seriously about how warbots will change warfare and what we can do to control them. The effects will be profound, and not limited to the battlefield. Consider—what will be the relationship between societies and armed forces when those forces consist of technicians and many thousands of robots, rather than pilots, sailors or soldiers? Who will be a warrior? Will societies be more willing to use force if their own people are less at risk?

If my forecast is right, the pressure to design warbots will outpace our ability to regulate them. The systems of today are sophisticated, albeit largely limited to situations where speed is of the essence—when missiles are flying and a defensive system must respond in an eye blink. What happens when warbots are more than last-ditch defensive systems? When they can manoeuvre, identify targets, track them and kill them—at great range, and with tremendous destructive power?

So, this book tells the story of AI that is at once remarkably smart—capable of inhuman feats of calculation, of sorting through masses of data and finding obscure connections—but also fantastically dumb. AI might not understand what we want, or the risks that we deem acceptable. And if not, it won't care a jot. It feels no emotion and is motivated only by the instructions we issue. It doesn't want to please us, and nor does it fear punishment if it transgresses.

The technology we'll encounter is real enough, or soon will be, even if some of it sounds bizarre. Later we'll explore the possibility of mind merges between human and machines, and of AI-enabled genetic engineering. Astonishingly, there's nothing here that's not entirely feasible. It's the connection of this advanced AI to war that is so troubling. And that is the likely, perhaps even inevitable, extension of the trends already underway.

This AI revolution started in a society that successfully melded the entrepreneurial spirit, scientific ingenuity and the deep pockets of the Federal government. That society celebrates individual freedom and has developed safeguards to protect it. But the revolution hasn't ended there—now China, Russia and other authoritarian governments are players too, all with a very different model of how AI research should be organised, and different ideas about the ends to which that research might be used. As the military competition heats up, the challenge for liberal democracies is to balance the risks to national security presented by China and Russia with the need to protect our own values.

Rules for warbots

All this is making many people uneasy: how can we be sure that AI will do what we want it to? Should it be able to decide who to kill on its own? How can we stop it if it goes wrong? Should we ban it? The activists of the Campaign against Killer Robots think so. They want warbots outlawed before they take to the battlefield. They have some powerful support. In 2017, Elon Musk headlined an open letter calling for a ban, joined by more than over a hundred AI experts, including some titans in the field. Together they warned that:

> Once developed, lethal autonomous weapons will permit armed conflict to be fought at a scale greater than ever, and at timescales faster than humans can comprehend. [...] We do not have long to act. Once this Pandora's box is opened, it will be hard to close.[4]

Unfortunately, the box is already open. Fully autonomous lethal weapons exist and even if they probably haven't killed anyone yet, it is just a matter of time before they do. Moreover, the benefits from warfighting AI are so great that many powerful states won't willingly sign up to regulation. They rightly fear that rivals might cheat, and so gain an insurmountable advantage if it ever came to war. Today, an AI arms race is underway. Moreover AI technologies are used in so many aspects of modern militaries that ringfencing it will be almost impossible. Pilotless drones are one thing, but what about an early warning system, or even the computer that sorts through pension details in the personnel back office? Even defining what constitutes AI is deeply problematic, because it's more a philosophical matter than a discrete set of technologies.

All this makes for dangerous times ahead. AI can behave in unexpected ways, and some of these will challenge longstanding ideas about war. Perhaps framing guidelines for how to use our own AI might be useful. We can't impose these on rival states, but they might themselves arrive at something similar—driven by the same uncertainties and anxieties.

Where to start? In the 1940s, Isaac Asimov crafted some famous rules for robots, which have been much commented on since. The title of this book borrows from his collection of short stories, *I Robot*.[5] Asimov's rules were intended to safeguard humans against danger from their machine creations. He used his stories to explore the concept of 'robo-psychology'—the idea that machines might have a mind of their own. Decades later, the Will Smith blockbuster movie of the same name explored similar ground—what would happen if a robot (the word means 'slave' or forced labourer in Czech) went rogue? No longer our slave, but a spirited individual with agency of its own. Asimov penned three rules that robots should follow:

- A robot may not injure a human being or, through inaction, allow a human being to come to harm.
- A robot must obey orders given it by human beings except where such orders would conflict with the First Law.
- A robot must protect its own existence as long as such protection does not conflict with the First or Second Law.

They may be good for science fiction, but Asimov's rules are a really poor guide for anyone planning to use AI and robots in war. After all, the whole point of a warbot *is* to kill. Warbots will do plenty of other things too—some of them a long way behind any front line. They'll perform tasks as boring and mundane as sorting the mail and working out which soldier might need a routine doctor's appointment. But they'll also be involved at the sharp end—combing through mountains of intelligence to figure out what enemies intend to do; advising commanders on the best course of action; assembling and dispatching fleets of lethal robots; and, ultimately, deciding who to kill. Violence, after all, is the distinguishing feature of warfare, and warbots will be extremely efficient killers—accurate and relentless.

So the rules are out, as they stand. But perhaps we can use them as a basis to think about how we actually want our warbots to behave. Rule One clearly needs amending—our warbots will be actively seeking to kill enemy combatants. And while they may seek to protect soldiers on their own side and any civilians caught up in the battlefield, we can't assume that. After all, human commanders are sometimes prepared to sacrifice troops in actions they know will be extremely hazardous, if that fits the wider purpose of war. Why would a machine give up that cold-blooded approach, if it works? Similarly, with civilians on the battlefield: commanders might not *want* to kill them, but what if they can reasonably anticipate that some will die in any assault? Is that enough to call off the attack? Not always.

What about Rule Two—that the warbot should follow orders? Following orders is precisely what you'd want your warbot to do: no one wants a rogue robot going berserk in battle, or one that has access to nuclear launch codes escalating wildly in a crisis. But it's not that simple. Whose orders should be followed? That seems a banal point, perhaps—of course, it should follow the commander; and ultimately, in a democracy, that would mean the President, or Prime Minister. But when? When they decide to launch it into combat, or during the action, as events unfold and the human leader changes his or her mind? What about afterwards? Perhaps the warbot should take a longer-term view of our wellbeing? If we make a rash decision in the heat of the moment that we will regret later, should our machines reflect on how the future version of us will see that?

Rule Three is out too. The Harpy is an existing Israeli loitering drone that can independently seek out and destroy targets. When it finds one, the drone turns into a bomb, diving in and exploding on the target. You might want your warbot to protect its existence in battle, but then again, you might want it to take risks to achieve the mission; risks that you'd be more reluctant to take if there were human lives involved.

Clearly designers of warbots will need different rules for their robots. But what, precisely? In this book, I hope to offer some answers. As an opening gambit, here are my rules for warbots:

- A warbot should only kill those I want it to, and it should do so as humanely as possible.
- A warbot should understand my intentions and work creatively to achieve them.
- A warbot should protect the humans on my side, sacrificing itself to do so—but not at the expense of the mission.

We'll revisit those at the end of the book, in the light of what we've learned about warbots along the way.

Warbots at the movies

Asimov's rules are still influential in AI debates today, seventy-five years on. Partly that's because popular understanding of AI has long drawn on science fiction. We can certainly learn something about AI warfare from popular culture, especially when thinking about the near future. But we need to be careful: the movies have created unrealistic expectations about AI.

Wilful robots and the efforts of humans to control them are both Hollywood staples. From the arty highbrow of Stanley Kubrick's *2001: A Space Odyssey* and Alex Garland's *Ex_Machina* to the knockabout fun of *Wall-e* and *Austin Powers*, the basic conceit is the same: Hollywood's AI is human-like, while being not quite human. Often these almost-human robots have malign intentions. All of which points to an interesting feature of many AI discussions—the connection between science fact and science fiction. Articles about AI in the popular media are frequently accompanied by a picture of Arnie as the *Terminator*, his robotic eyes glowing a sinister shade of red. This infuriates some AI specialists, who argue that AI like this is so far distant as to be unfeasible, whereas 'killer robots' are a clear and present threat. Perhaps it doesn't help that the British Ministry of Defence has called its network of communications satellites 'Skynet', also the name of the artificial intelligence system that attacks humanity in the *Terminator* films.

Arguably, these films tell us rather more about humans than machines—starting with our tendency to anthropomorphise; to compare AI to human intelligence. The AI in movies looks rather more like a human than it does the AI we have today. Our AI plays chess like a grandmaster, but you don't have to know very much about how it works to appreciate that we're a very long way from the sinister Agent Smith of *The Matrix* and the Replicant hunting Deckard of *Blade Runner*.

Still, perhaps we can learn something useful from Hollywood about warbots. First, quite simply, that AI has a strong hold on the public imagination—decades of hit movies are testament to our fascination with machines that think. Second, that these films are really about human stories, not machines. When filmgoers watch AI, they aren't just thinking about what makes the machines tick—they're thinking about what makes us human. Our fictional robots are human-like, but not quite. Sometimes they struggle with an emotional awakening, a flowering of consciousness that is rejected by humans—that's the fate of Rutger Hauer's doomed 'Replicant' Roy in *Blade Runner*, reminiscing sorrowfully about 'attack ships on fire off the shoulder of Orion'. Or they lack an appropriate emotional response—as with the coolly manipulative Ava, of *Ex_Machina*, and Scarlett Johansson as the enigmatic girlfriend in *Her*. It's a well-worn trope. Hollywood's AI sometimes has emotions, and often it's self-aware. It might be a warped sort of human, even a psychopath, but it still looks like us, seen through a glass, darkly.

Third, the movies feed into unrealistic expectations of what's possible with AI. The weakest scene by far in *Ex_Machina* is the one where the mad genius explains how Ava works, while holding a gelatinous and faintly glowing artificial brain. Is that really what a sentient machine brain will look like? It's unlikely that anyone watches the film for a primer on AI hardware, and in fairness there's already a huge amount of philosophy of mind smuggled into *Ex_Machina's* brilliant script by the renowned AI scientist Murray Shanahan. But machine minds like Ava's are very far distant from the near future of the film's setting, so perhaps it's only proper that Ava's brain is unconvincing.

Fourth, in Hollywood, robots go wrong, usually with disastrous consequences. After all, that's the essence of drama. Better yet if the disaster is unforeseen by all except a wise visionary, who was distrustful of machines all along. Robots can certainly redeem themselves in this fable, usually by demonstrating their human qualities—like empathy, or loyalty. That's what the cyborg Bishop does, nobly battling the reptile *Aliens* to save the heroine Ripley. In the real world, there's plenty of scope for AI to go wrong too—that's what fuels the concern of campaigners against 'killer robots'. But in the real world the machines don't go wrong because they develop a free will, or because they're psychopathic. They go wrong because they're unintelligent, at least in so far as intelligence would involve correctly understanding what we want them to do. Occasionally that happens in Hollywood too, to be fair. In the 1987 film *Robocop*, the ED-209 robot makes a catastrophic error. During its demonstration to city authorities the machine mows down a hapless corporate flunky, having first instructed him to put down his weapon. This he does, but the robot doesn't sense it, perhaps because the sound is muffled by a carpet.

And lastly, this being Hollywood, there's often a great deal of bloody violence involved. All the way from blindingly fast knife-play by Bishop to the prospect of thermonuclear war in the classic *WarGames*. That 1983 film features an AI called WOPR, capable of launching a nuclear strike against the Soviet Union. The Pentagon, we learn, has introduced an AI system because human operators proved reluctant to launch nuclear strikes during an exercise. Thankfully, when it comes to it, the computer seems similarly reticent. When the movie's young hacker inadvertently gets into a confrontation with it, WOPR messages him, 'what a strange game, the only winning move is not to play'. Instead it proposes a nice game of chess. In the real world, rumours abounded of a Cold War computer system from precisely this era which automated nuclear retaliation. This was the Soviet 'dead hand' system—designed to respond to a devastating nuclear attack on the USSR that left no human capable of deciding what to do. At a time when computers were rather less powerful than today, especially in the Communist World, that was an alarming prospect. But it turned out the computer was really a reinforced bunker, deep under Moscow, staffed with humans on whom that onerous responsibility would fall.[6]

Yes, the relationship between Hollywood and warbots is tenuous at best. It's only fiction after all. Still, movies point to some powerful ideas that recur throughout the history of AI. We'll see some when we get to the chapter on AI strategy, where machines offer insights at higher levels of war than merely being robot gunslingers. From the brilliant *Gattaca* comes the notion of genetically modified humans with enhanced cognitive capabilities (and status to match). It is, we'll see, increasingly feasible. And from *The Matrix* comes the idea of a hugely detailed simulation of reality, in which we are unwittingly living. Right now, the UK's Ministry of Defence is working on 'Project Improbable', a richly detailed synthetic environment—complete with artificial cities, populated by artificial people. What if you could run a future operation in the simulator first, to see how it pans out? Is it a good idea to blow up the power grid? What sort of casualties might be caused by a bombing raid on an enemy building? If it's sufficiently realistic, a synthetic environment like this would be a good place to rehearse real operations or exercise military commanders as part of wargames. It might even be a good place to think about the design of future warbots.

It's not just movies either. In his 'Culture' novels, Iain M. Banks imagines a universe where an advanced civilisation has created super-intelligent AI.[7] Human and machine work as a team to fight against their alien enemies. The artificially intelligent warships have their own personalities and give themselves witty names to match: *GSV So Much for Subtlety*. Our AI is less amusing, but the prospect of humans and intelligent machines teaming up to fight war is already here. In battle, air forces are working on robot 'loyal wingmen' to fly alongside human pilots;

back at headquarters, they're developing AIs to analyse information and suggest plans alongside human staff officers. The humans of the Culture novels and those developing AI weapons today have a common dilemma—how much trust to put in the judgment of machines.

For now, that's probably enough fiction. We can learn a lot about warbots from the movies, and some of these themes recur in the chapters to come. But there are limits to the vision of even the best Hollywood writer. To see that sci-fi is a prisoner of its own era we need only watch the disco-inspired *Buck Rogers*.

Warbot psychology

Asimov's rules prompt interesting questions about robot minds. Marvin Minsky, one of the legendary pioneers of AI research said that reading Asimov left him astounded and unable ever after to stop thinking about how machine minds might work.[8] It's the minds of warbots that concern us here. In thinking about war, our attention is inevitably drawn to weapons—the latest aircraft, missile or submarine. But AI is more than a weapon. When it comes to understanding Artificially Intelligent conflict, our focus should be squarely on the mind of the machine, not its engines or cannon. If machines are deciding about violence on our behalf, we should be thinking very carefully about how they'll do it. We need to know how they will interpret our goals and what assumptions they'll make about the enemy. To date, however, there have been vanishingly few attempts to think about AI strategy.

Warbots' brains will certainly be very different from human ones. Much of this book explores the differences. We know that human warfare is an intensely psychological affair; the question is how that changes with the introduction of AI. Carl von Clausewitz is perhaps the greatest scholar of war. In his master work, *On War*, completed by his wife and colleague Marie after his death, Clausewitz made many astute psychological observations.[9] It's these insights that have given his writing a timeless value, elevating it above the historical particulars of warfare in his day.

Clausewitz knew that war involved intense emotions—he wrote, for example, about the 'passionate hatreds' that motivated the belligerents. And he understood that war involved decision-making in difficult circumstances. He described an idealised 'genius'—a commander who would make the right decisions amidst the press of events, even with limited and unreliable information to hand. Clausewitz conceded that he was in the dark about how much of this actually worked. Psychology was in its infancy as a science—still very much the fodder of armchair philosophical discussion. But he knew, at least, that the brain didn't look much like a machine. The mathematically inspired strategic theories of some contemporaries left him cold—like many so-called 'Romantic' thinkers of the day, he preferred to think about more ephemeral forces at work. A 'ghost in the machine', as it were.

In the decades since then, scientific research, especially in psychology and neuroscience, has advanced our understanding of human decision-making dramatically. And we know rather more about warfare too. From archaeology, primatology and anthropology we have gleaned insights about how other people and even other species fight. Today, we have a decent grasp of the ways in which human decision-making shapes conflict—of why we fight, and how. In fact, some argue there's an inter-relationship between the human mind and warfare—in the insecure and sometimes violent epoch of human evolution, warfare shaped the human mind, just as assuredly as the mind shaped warfare.

Artificially intelligent warfare will be quite different. Because it is a decision-making technology, AI is qualitatively different from other technologies that have shaped war—from stone hand axes to thermonuclear warheads and everything in between. Those were tools, whereas AI is an agent. This cognitive aspect is central to thinking about AI in war. Take emotions. Humans use emotions to guide their behaviour, and that includes behaviour in combat. Machines don't have emotions, at least not yet. They'll decide how to respond to their adversaries and how to coordinate with their allies using very different techniques from ours. That means different tactics in battle, but also the prospect for larger differences in war.

And then there's creativity. Humans have it in spades. The incumbent world chess champion Magnus Carlsen is a great example of human creativity—a mix of natural genius (for memory, and pattern recognition), but with something extra. Carlsen argues that this is intuition, an instinct for the right move. But what is intuition, and what makes Carlsen's instincts better than mine? And why are the world's best chess players vulnerable to number crunching machines that don't use intuition at all? Might the most creative generals be similarly vulnerable?

Clausewitz eschewed chess as the best metaphor for war. He thought it was more like poker. There, rival players each have a different and partial view of the game—seeing only the cards they hold. And poker involves luck too, much like war. As it happens, though, AI can now beat the best poker players in the world. But Clausewitz wasn't thinking about machines. He thought that his genius commander would have an instinct for war—one that might be honed by experience and reading. War, he reckoned, was more art than science—an intensely human and creative phenomenon.

These insights are what makes Clausewitz a timeless genius. Today it is cool to critique him, to suggest that he is somehow irrelevant for the times. That is certainly not my intention here. However, warbots will undoubtedly change the psychological essence of warfare. Human intelligence won't disappear from the battlefield for many years to come, if ever. But for the first time in history, a new intelligence is going to war. We would all be well advised to think very carefully about the minds of warbots.

HOW TO BUILD A WARBOT

There's no single blueprint for warbots, and a bewildering array of possibilities. We are currently in an experimental phase roughly analogous to that of piloted flight in the early years of the Great War, when a variety of eccentric looking aircraft took to the skies to do battle over the trenches of the Western Front. We can learn something about warbot design from that era, particularly about how function shapes form. In only a few short years of conflict, the core air power roles that remain central to modern air forces were established. It turns out that there are only a handful of things you can use a military aircraft for.

The first use of air power was for reconnaissance, and long predates heavier-than-air flight: Revolutionary France formed an 'aerostatic corps' and its balloons were used for the first time in battle at Fleurus in 1794. Gaining altitude was a good way of increasing battlefield observation and also afforded the scope for ground attack—the next role to be established. In 1911, Giulio Gavotti, an Italian flier, was apparently the first to demonstrate this new role, dropping a grenade from his rickety aircraft on doubtless startled Ottoman troops below. Next came the struggle for control of the air, as ground-based guns blasted away at overflying spotter aircraft. Soon the planes themselves were engaged in combat with one another, spurring a rapid burst of technological development, for example of machine guns that could fire through the propeller. The last core air power role, that of moving things around, took a little longer to emerge, until aircraft became larger and more robust.

So perhaps a good starting point is to ask what we want our warbot to do and work backwards from there. Unfortunately function alone can only tell us so much about the characteristics of our warbot. Consider, by way of example, the Lockheed SR-71 Blackbird. Aerial reconnaissance may well have been established with the Aerostatic Corps, but a balloon floating elegantly over the battlefield is very different from an SR-71 Blackbird. This graceful aeroplane came into service in 1966 but still looks like technology from an advanced alien civilisation. The Blackbird was astonishing. Capable of speeds in excess of Mach 3, and cruising in the upper atmosphere at 85,000 feet, it could fly faster and higher than the best Soviet piloted interceptors. At three times the height of Everest, its pilots needed pressurised spacesuits. Ground-based anti-aircraft missiles were faster than the SR-71, at least in theory, but by the time they reached its altitude they would be almost spent. Vietnamese Communists and their Russian technicians fired several hundred missiles at Blackbirds during the Vietnam war, entirely without success. The technology was exquisite. It flew so fast that heat generated by friction was so intense that the panels of the airframe would expand—at subsonic speeds it leaked fluid through its open joints. The Blackbird's sleek curves enhanced its aerodynamics while reducing the radar cross section—an early indicator of stealthy aircraft to come. A technological wonder, the remarkable SR-71 first took flight only fifty-eight years after the Wright brothers took to the skies near Kittyhawk in 1903.

I mention all this to highlight the rapid pace of technological change in completing essentially the same task. The Blackbird itself was obsolete within fifteen years. Ruinously expensive, even for the US, it was also risky to operate. What if one malfunctioned over enemy territory, or one of the enemy missiles managed to hit it? The SR-71 didn't fly over Soviet airspace for precisely that reason. Moreover, there was a safer, uncrewed option; similar intelligence could be gleaned from increasingly sophisticated spy satellites, operating with impunity from space. These were hard to re-task—that is, to give new missions outside their original objectives, especially where that would require a new orbit. But they were soon abundantly available, allowing ever greater coverage. And soon they would be joined in the reconnaissance role by pilotless drones, including those able to operate stealthily in defended airspace, and then later by space vehicles that could easily be repositioned, like the experimental X-37, which looks like a miniature space shuttle. From Gavotti's flimsy biplane, through the space-age Blackbird, and on to the shadowy RQ-170 'Beast of Kandahar', a pilotless drone used to track Osama bin Laden in his Abbottabad hideout—all that in only a few decades. The pace of technological change is accelerating.

Another reason that function might not be the best way to think about warbots is that technology can be transformative, not just evolutionary. In the apocryphal tale, Henry Ford captured the idea well, noting that if he'd asked potential customers of his cars what they wanted they'd have asked for a faster horse. As Ford knew, thinking about cars as a faster horse rather misses the transformative nature of the automobile: it can do things a horse simply cannot. The same is true of the warbot—it might make for a better tank, but then why make a tank at all? Likewise aircraft—will our warbot be better at performing those core air power roles, established in the early twentieth century, or will it do something else too? Does it even make sense to think in terms of 'aircraft'? Perhaps the same warbot will operate in the air and underwater, like a gannet diving for sardines. We can't just start with the function and work backwards. We also need to work forward from a deeper appreciation of the technology to understand what it makes possible, and then we need to see how that changes the big picture of war.

Still, it's worth keeping one large goal at the back of our mind. For Ford's blinkered customers, looking for a new jalopy, the objective was to get from A to B, safely, reliably and, ideally, speedily. For the military in war, they are tasked with employing violence in order to achieve some political aim of the society that sent them to fight. That is the essence of strategy—using violence for a larger purpose.

Lastly, function and available technology aren't the only drivers shaping the sort of weapons that armed forces acquire. Among other factors in play are established organisational and societal attitudes to war. Scholars talk about 'strategic culture' as a factor shaping the way groups think about war, including the sorts of technologies they employ. As we go along, we'll see that military and societal cultures are likely to play key roles in the development of warbots. We'll also discover that the direction of travel isn't just one way—the technologies that become available will in turn influence cultural attitudes about war.

The warbot brain

Another way to think about warbot design is to consider what lies inside. Specifically, to ask what sort of artificial brain will allow the machine to operate effectively, in accordance with our strategic goals and with my proposed rules for warbots. Will the machine need to be clever? What does it even *mean* to be clever? This is a question that lies at the heart of research in Artificial Intelligence. Less obviously, there's long been a link between defence and those thinking about that question. Much of the research in AI has been driven (and funded) by organisations that are interested in using the product of the research for military purposes. It still is.

On one view, the link between war and intelligence goes far deeper: the connection between intelligence and fighting stretches right back through evolutionary history.[1] Violence shaped brains, for humans and other species alike. Now, in the computer age, the link persists. Computers themselves were developed in wartime, and those charged with national security were quick to grasp the military implications of machines that could act intelligently. Code-breaking, translation, imagery analysis, to say nothing of autonomous weapons—all were immediately seized upon as ripe for research.

But before we get to all that, it's worth exploring in more detail what constitutes intelligence. There is no single settled definition, just as there is no settled definition of Artificial Intelligence. But there are some common themes in the voluminous literature on AI, and these might serve as a useful jumping off point.

Many theorists focus on information processing as the distinctive feature of intelligence. But to what end? The purpose of intelligence is frequently considered as being to do something useful—though, in fact, the ultimate purpose of many acts by intelligent humans is often obscure. Another element is the capacity to learn, that is to improve the processing of information somehow. Learning is particularly useful where it produces a change in behaviour, so that it is better suited for the goal of the intelligent actor. Intelligence seen this way is about responding to the environment. Learning is the hallmark of human intelligence—the ability flexibly to respond to change. A final distinction to consider: learning may be acquired through experience, or it may be more 'hardwired' into the agent. For humans and other living organisms, evolution is the driving force there—the struggle to reproduce selects the most adapted intelligence.

All these ingredients would be part of my definition of intelligence. It's the ability to sense the environment and respond to it in ways that are useful for some goal. This is a sufficiently broad definition to allow us to apply it to agents or systems that are not alive, like computers.

Human intelligence is intimately connected to our bodies. Our brains evolved to allow our bodies to navigate an environment filled with threats and opportunities. That environment is intensely social: we depend on our group to survive, so we have evolved a socially attuned intelligence aimed at working out what others are thinking.[2] Our intelligence is also emotionally informed—it feels like something to be us.[3] Emotions are a way of ordering our cognitions, of deciding what's important. Emotions connect our biological motivations with our behaviours. Our emotional experience involves consciousness too, of course. That phenomenon continues to baffle science, but one plausible rationale is that consciousness is simply a way of telling a coherent story, to ourselves and others, about what we are doing in the world. And there's one last important element which follows from that—our human intelligence is driven to understand, and to find meaning.[4] We don't just want to know *what*, but also *why*. It's this drive for meaning that underpins our cognitive flexibility. It shapes our curiosity and fuels our creativity. And it allows us to respond appropriately in novel circumstances, with limited experience. In short, human intelligence, as the jargon has it, reflects our ontology: our experience of what is real in the world derives from our existence as embodied primates.

Machine intelligence looks very different. For one, the goals for which the intelligence works are not intrinsic: machines want what we tell them to want. They have no comparable biological urge to survive and reproduce. Much flows from that fundamental distinction. The way in which machine bodies shape intelligence is different, so too are the means by which machines navigate their social environments. There's no emotional aspect to machine life, and no conscious experience either. And the search for meaning and understanding looks very different for a machine. Douglas Adams playfully captured the difference by having the superintelligent computer Deep Thought answer '42' when asked for the meaning of life in *The Hitchhiker's Guide to the Galaxy*.[5] Playfulness, creativity, imagination and curiosity are terms we associate with human intelligence, not machines. There may be AI facsimiles for all these things. After all, machine intelligences can be embodied and act in a social context, so there are similar rationales. But they will still be rather different phenomena—in scope and in underlying processes. Like ours, machine intelligence reflects the underlying ontology of the machine.

This all matters when thinking about warbots. What will they want and how will they go about getting it? How will they interact, and how will they understand our intentions? The answers will inform how we think about AI in war—both in the heat of battle, and on the larger canvas where strategists seek to employ force to achieve our collective goals. Just as human intelligence was shaped by the prospect of conflict with other humans, so we can expect war to play a part in shaping machine minds. It already has.

The Artificial Intelligence of today may seem incredible—and on many measures it is superhuman. In recent years, world class players of chess, Go and even poker have all been soundly thrashed by modern, cutting-edge AI. Mastery of those games has long been assumed to require high levels of intelligence, with excellent judgment of the odds, prodigious memory and creativity. And now machines can outperform the best humans. But if modern AI is so brilliant, why is the Siri personal assistant on my iPhone so dumb? It can return train times, or set my alarm clock, but anything that requires deeper understanding seems entirely beyond it. Clearly something is missing from even the best modern AI. Too often AI lacks the sort of intelligence that humans take largely for granted. Can it, for example, converse convincingly with a human? Can it write a story, compose a tune or paint a brilliant work of art? These are the sorts of skills that might persuade us it had human-like intelligence.

Conversely, perhaps those are the wrong sort of skills to be thinking about when gauging intelligence. We certainly have a tendency to judge by human-centric standards. Perhaps we shouldn't. A tell-tale is the term 'strong AI', used by afficionados to mean AI that can perform like a human—flexibly, socially, emotionally. But that's certainly not the only yardstick for intelligence. After all, we may be very good at those things, but there are plenty of other things we are poor at. We are for example weak at statistical reasoning and our memory is patchy and selective. Machines excel at both skills.

Fairer to say then, that the sort of intelligence that AI can do right now is limited. It is often described as 'narrow' in the literature—good at one task, or a small set of skills, and markedly less flexible than ours. But even so, it's definitely good enough to achieve high, superhuman performance in some areas, including areas that were previously thought to require human insight. And it's certainly good enough for some radical new forms of warfighting technology. In fact, even this limited sort of intelligence is capable of transforming war, bringing changes more radical than those wrought by earlier technologies, even nuclear weapons.

Autonomy is not intelligence

The signature feature of a warbot is autonomy—its ability to operate without moment-to-moment control by humans. This is somewhat distinct from intelligence, because an agent can be supremely unintelligent and still act autonomously. It may follow a pre-ordained series of steps without intervention, and yet be entirely unresponsive to changes around it. That's a rather limited, wholly automatic autonomy which will be poorly equipped to deal with novelty. But it would still allow the execution of its mission in favourable circumstances. A richer, more intelligent form of autonomy is able to respond flexibly and more creatively to change—perhaps learning from experience, or through observation of others.

To illustrate these distinctions, consider the humble, but very successful (in evolutionary terms) ant.[6] Each individual ant has limited cognitive ability, certainly when compared to humans. The ant responds to its environment in ways that are predictable and instinctive. Its intelligence is innate and robotic. It is intelligent, because it can change its behaviour in response to events around it. But there are limits—the ant's learning has largely been done by evolution, and thereafter there is limited capacity to improve performance itself. It can, for example navigate by a sense of smell, or by sensing polarised light (in the case of desert dwelling ants, for whom shifting sand renders smell problematic). Change the scent trail or the light, and the ant follows along. Intelligence researchers sometimes call this sort of cognition 'situationalist'. The ant's actions are automatic, instinctive and rather inflexible. They're largely 'built in'—the ant comes pre-equipped to respond, it doesn't reflect or learn to improve its behaviour, from direct experience or watching other ants.

Still, that's often good enough. The ant is one of nature's most wildly successful creatures. Added together, the biomass of ants far exceeds that of humans. Like us, ants have spread around the world, surviving in a range of hugely diverse ecosystems. Different species have evolved markedly different approaches to life, suited to their own environmental niche. Sometimes these are very large niches—the biggest single colony of ants yet discovered stretches some 3,500 miles across southern Europe.

There is extraordinary diversity here, and also extraordinary complexity. My own personal favourite is a species whose queen has adapted her body to fit astride the queen of another species—via a concave thorax and the ability to cloak her pheromones to avoid detection by the guardian ants of the unsuspecting host. Evolution can produce extraordinary complexity.

But there are dangers to this degree of automaticity. Change the niche, and there's a real risk that automatic behaviour becomes maladaptive. The dodo famously did fine until humans arrived with non-native animals to disturb its natural habitat. Animals are all prisoners, to some degree of their environmental niche—including humans. But we do rather better than ants at adapting to change. Partly this is physical: we have bodies that can cope with a wide variety of environmental conditions. We can thrive at high altitudes, in the frozen arctic tundra, the scorching Saharan desert and the dense rainforests of the Congo basin. But the main reason for our success is cognitive. We can do all this because we possess tremendously flexible brains, far beyond that of the ant. Among our most useful cognitive skills are our intense sociability and our capacity to learn (and teach). The two are related—we learn from others, producing rich human culture, unique to our species. We learn cumulatively, building on the achievements of others by understanding what they are trying to do. Moreover, we have the ability to generalise our learning from one task to another, adapting our skills to suit new conditions. Language helps immensely—giving us the ability to explain complex ideas and share knowledge, even when the object isn't physically present. All this is predicated on our 'theory of mind'—the empathic ability, partly instinctive, partly the product of deliberate reasoning, to think and feel what others are thinking and feeling.

Lest we get too complacent about the glory of human innovation and intuition, we ought not to forget the ways in which we, like ants, make our decisions instinctively, on the basis of hardwired, situationalist thinking. See a fast-moving snake in the corner of your peripheral vision, and feel a surge of fear, even as you bolt for safety. So, we too are also constrained to a considerable degree by the situation in which we find ourselves. Large parts of our cognition work along similar lines to the ant—automatically responding to sensory inputs in ways that have been hardwired into us by evolution. But it's the parts that are distinctively human that have allowed us to dominate environmental niches worldwide.

Ants are agents

There's an interesting philosophical question here—if its behaviour is fully automatic, and is entirely, or almost entirely, determined by the situation, does it make sense to talk about an entity as an 'agent'? Is it really behaving with autonomy, or is its behaviour entirely determined by outside forces, sensed, and then mechanistically processed into actions? For all their automaticity, it makes sense on one level to consider the ant as an agent, however minimally. An individual ant is alive, it has goals, and within its capacity to sense and react, it makes choices.

Our biology is one fundamental difference between us and the sort of artificial minds that we might imagine doing the decision-making for our warbot. The ant, like us, is engaged in a struggle against entropy. This concept, fundamental in physics, tells us that energy dissipates, and that disorder progressively overtakes order. On a cosmic scale, it's been that way ever since the big bang. Energy was unleashed, since when it's been expanding and cooling. In all the cosmos, so far as we know, there's only one thing pushing back against the forces of entropy—life. It's the big difference between me and the intelligence of the increasingly battered laptop on which I'm writing this book. Life is the self-organisation of matter, and its onward organisation through reproduction.

For human and ant alike, there is a fundamental biological imperative driving our behaviours. Look down at the tiny ant, busily working away according to its limited menu of options. Does it want to survive, as (most) of us do? Perhaps its incentive to survive individually is less than ours, since what matters most is the survival of the colony, via its queen. Nonetheless, like us, the ant's behaviour is geared to enhance the onward journey of its DNA into successor generations of ants.

Warbot ants

Machines lack the biological motivation that would signify agency. But we can still consider them agents, in the sense that they choose. They don't choose their own goals, which constrains their autonomy. But then the ant and I have both inherited our master-goals too—survive, reproduce. Within the constraints of that inheritance, machines, like ants, have some capacity to decide. My question about agency is really about free-will, which along with consciousness is one of the great unresolved topics for those interested in philosophy of mind. We won't resolve it here either. But insofar as any of us is *really* choosing, rather than simply responding in a pre-determined way to our environment, we can think of us all as agents: ant, human and computer. Machine agency, like machine intelligence, may be qualitatively different from human—resting on very different foundations. But it is agency nonetheless.

Situationalist behaviour like the ant's offers a useful way to think about building our warbot. Starting with simple behaviours and even simpler cognition is easier than setting out to build a superintelligence from scratch. And that's exactly what early researchers did, especially those working in the development of autonomous weapons.[7] They lacked the sort of computer power that would be capable of doing anything differently, so all they could do was build something that could carry out its behaviour without human input, and perhaps even respond in some fashion to its environment.

The V1 bomb developed by the Germans towards the end of the Second World War is a great example of this sort of dumb automaticity. A primitive cruise missile, the bomb used a small pulsejet engine, and set out on a pre-determined path. Down the ramp it went and zoomed off over the channel from the French and Dutch coastline, towards London. A simple gyroscope autopilot regulated altitude and airspeed, while an odometer driven by a small anemometer on the nose counted down the distance remaining to the target. Good enough for area bombing, but not very precise. When the count reached zero, the V1 transformed into a dumb, or unguided bomb, falling into a dive. The engine cut out, its incessant buzzing replaced with an eerie silence. The 'doodlebug', as Londoners christened it with typical *sang froid*, claimed hundreds of lives, but proved not to be the devasting, war-winning weapon that hard-pressed Germany needed. For one thing, the amount of explosive was limited by the size of the airframe. For another, the British quickly fielded an effective countermeasure—radar guided anti-aircraft batteries, of which more shortly.

Basic or not, the V1 was revolutionary. Some 8,000 were launched against Britain, and while the total tonnage of explosives was only a quarter of that in the Blitz, earlier in the war, the raids inflicted casualties at the same rate, and the Germans suffered no casualties of their own (versus more than 7,500 in the Blitz).[8] Initially, the designers contemplated fielding the V1 as a radio-controlled device, which would have allowed more precision. But why bother if it was intended primarily as a weapon of terror and revenge against city-dwellers?

Instead, the V1 amply demonstrated that autonomous munitions could work. Artificially intelligent? Not really—the gyroscope and odometer were mechanical, a product of the industrial, machine age, not the electronic era to come. But rather more importantly than the material of its construction, the V1 was not capable of responding to change in its environment. The bomb may have been autonomous, but it was pretty dumb. It followed the steps its designers intended, and very effectively too, but it could not adapt to overcome obstacles in its path. In addition to artillery fire, hundreds of V1s were dispatched by crewed interceptor aircraft. A few pilots even managed the spectacular feat of tipping the bombs off course by flipping its wingtips into an uncontrollable dive. Static barrage balloons claimed a few hundred more. Evasive action was not possible, because the V1 was an automatic warbot, but not an intelligent one.

This sort of dumb autonomy is found in other forms of early warbot—the proximity fuse that trips a maritime mine when a metallic hull passes nearby is a good example. The mine can be left on its own for months at a time and still detonate entirely without human supervision. The same automaticity is found in depth charges, which are set to explode when triggered by a pressure sensor. Or in even less sophisticated improvised explosive devices and booby traps—a mechanical clock used as a timer that forms an electrical connection when the hands meet, for example. Weapons like these can be devastatingly effective, but are they agentic? Not really—they are passively carrying out their instructions. And short of malfunction, they carry them out to the letter, unless some form of novelty intervenes. Still, they serve as a reminder that autonomous warbots need not look like those in our imagination, whether inspired by Hollywood's flights of fancy, or by our experience with the digital computers of today.

Slightly more sophisticated warbots can respond to changes in their environment, a basic ingredient in intelligence. A homing torpedo can close in on the sound made by a ship's screw churning the water; an air-to-air missile manoeuvres to chase the heat signature emitted from the engine of an enemy fighter jet. There's a human involved in firing the weapon, just like the depth charge, and the V1. Just as there's a human involved in laying the anti-personnel mine, or priming the timer fuse of an explosive device. In modern jargon, this is referred to as a 'human in the loop'—that is, the decision-making loop. The machinery can't make the decision to kill, but once set in train, there's no need for direct control. And (unlike the depth charge, anti-personnel mine or the IED) homing torpedoes and heat-seeking missiles can adapt, changing course to close on their target. It's still a pretty basic form of intelligence, and rather limited. That creates vulnerabilities, for example, deception by dummy targets—whether the bright 'chaff' fired off by the defensive systems of a fighter jet, or the acoustic decoys released by a submarine under attack.

Like the V1, these autonomous weapons have a long history. The German navy introduced acoustic torpedoes in 1943, and the Allies had a countermeasure decoy in service later that year. Infrared heat-seeking missiles took a little longer. Again, the inventive German military was to the fore, but the Third Reich was defeated before completing a working system. In Britain, IR guidance lagged behind research on radar—used to telling effect to direct fighters during the Battle of Britain and later in guiding artillery fire against those incoming V1s. But by the late 1950s, autonomous heat-seeking missiles were in service with western air forces. Both weapons types fit the moniker 'fire and forget'; the human operator, not the machine, makes the decision to engage, but after that, the weapon system closes on the target, changing course if necessary, entirely self-sufficiently.

Consider again the radar-guided counter-battery fire levelled against the incoming V1. This was a complicated problem, since the gunners would need to gauge not just where the target was, but where it would be by the time the shell arrived. The solution employed a combination of microwave radar and analogue computer to track targets, alongside shells with an onboard proximity fuse that employed a short-range doppler radar. Each of these technical accomplishments was astonishing by itself, and their integration into an effective weapon system no less remarkable. It arrived just in the nick of time, dramatically improving the accuracy of defensive batteries, so that in one day late in 1944 more than 80 percent of incoming V1s were destroyed by flak. Much later, in the 1970s, the automated system grew more complex still. By then, more sophisticated digital computers would allow counter-battery fire, able to track the elliptical trajectory of incoming shells, work out their origin and compute a fire solution. Now, artillery would need to fire and then move, or else be hardened, in order to avoid devastatingly accurate return fire. All that lay ahead, but already by the Second World War the concept was a game-changer in battle. Automated warbots attacked, and automated warbots defended. And while humans were 'in the loop', with the power to initiate action, once the shooting was underway, there was little they could do. And all this before electronic and digital computing.

The wartime problem of anti-missile gunnery inspired one of the great early thinkers of Artificial Intelligence, Norbert Wiener. Wiener was fascinated by the concept of feedback loops, the central idea in his concept of 'cybernetics'.[9] This was a larger idea than computing and larger even than AI. Once you started seeing feedback loops in complex systems, you could see them everywhere—especially in biology, but also in sociology. Cybernetics is a useful way of thinking about situationalist intelligence. The V1 has none. If the environment changes, as an RAF Tempest fighter comes alongside to try and tip it out of flight, the V1 is wholly unresponsive. An incoming artillery shell produces the same result. There was simply no way for the doodlebug to sense, let alone respond to the incoming information. But the acoustic torpedo and the proximity fuse do have limited intelligence. They can sense what's happening in their environment, at least partially, and respond to changes in a way that is productive. Primitive, perhaps, but here are the roots of autonomous intelligent weapons. More advanced AI weapons would have far greater ability to sense, and then to make sense of their environment. They would be able to conduct more sophisticated decision-making on the basis of that information. And, finally, they would be able to engage in a suite of behaviours to implement their decision. That's what it means to be an agent—to be able, to some degree, to respond to a situation, in pursuit of some objective.

For the cybernetician, the agent's behaviour isn't the end of the matter—it prompts further changes in the environment, which in turn are sensed by the weapon, and so on, in a cycle of interaction and feedback. For the agent, it requires modelling the world, in order to see how it changes in response to the agent's own moves. That's fiendishly complex, and Wiener's book is awash with complex and impenetrable mathematics. For a while, there was a fashionable buzz about cybernetics, perhaps a bit like the one about 'deep learning' today. But neither the maths, nor the computing power of the day was up to it.

One early example of cybernetics that caught the public imagination was W. Ross Ashby's homeostatic robot, developed in the 1940s from surplus RAF bomb parts. There wasn't much intelligence going on here—when disturbed, Ashby's mysterious homeostat, all magnetic coils and resistors, automatically reverted to its original state. His paper 'design for a brain' was somewhat overstating things.[10] Still, homeostasis is certainly a feature of biological life. And insofar as it demonstrates motivated response to novelty, the homeostat was intelligent. Homeostasis is largely automatic in humans, as when the body sweats or shivers to maintain its core temperature. There's a desire in life to keep on keeping on. But cybernetics as a discipline is agnostic about the goal of all those feedback loops—certainly the heat-seeking missile has no desire to preserve itself.

Two final observations on cybernetics from these pioneering days of autonomous weapons. First, all this interaction suggests the need for careful thinking about what the boundaries of the agent are. What is an agent? It seems an innocuous question. The warbot is an agent, because it responds to the environment to achieve its mission. Surely a cruise missile itself is an agent? But what about the gyroscopes or GPS transponders that keep the missile aligned and on course? Aren't they agents too, in their own right? Their individual behaviour is what's responding to the environment. But then what about drawing the circle wider—surely the agent is the larger system that produces the effect (in this case the missile hitting its target). That would include the humans who feed the objective into the guidance system, who fuel the missile for its flight, or who steer the submarine carrying it within range of the target? These are all interconnected in a cybernetic fashion, with loops of action and feedback. The Portuguese Man o' War offers a good example from biology—it looks rather like a jellyfish, but is in fact an assembly of closely cooperating, but genetically distinct organisms, not a single animal.

Rather than individual weapons, then, it makes sense to think of weapon *systems*. So radar was a powerful, even battle winning technology for the RAF ahead of the Battle of Britain, in the summer of 1940. But the radar sets themselves were organised into a complex and integrated air defence system. The chain of radar stations along the coastline fed their information through to a centralised command and control centre at RAF Uxbridge. These were then integrated with sightings by observers and aircrew, with enemy tracks plotted on a large map. Commanders would then coordinate the deployment of interceptor aircraft from bases throughout south east England, weighing up, among other things, the need to keep aircraft in reserve, and the need for crews to rest and refuel. Radar itself was a vitally important technology, but success depended on the entire system, operating as an organic whole.

A second observation: cybernetics is a useful concept for thinking about war itself, seen as the violent and dynamic interaction of many variables. The role of the strategist is to think about the consequences of military action. The question they should ask is: if I do this, how will the enemy respond? Sometimes even expert strategists pay too little attention to that sort of question. In a longstanding study of political judgment, the psychologist Philip Tetlock discovered that subject-matter expertise was of very, very limited value in predicting what would happen next.[11] You might think that an expert in nuclear weapons, or in Iranian security would be better able than anyone to gauge whether Iran will acquire a nuclear weapon, and if so, what Israel or the US might do about it. But talk is cheap. Tetlock asked many experts for predictions like that, and compared them to what actually happened. A more useful sort of expertise than the subject-matter expert, Tetlock argued, was the generalist, who knew a little bit about a lot of things. More useful still, he discovered, were experts in decision-making, who understood the shortcomings of human judgment, especially judgments about the future. A large part of that expertise consists of updating probabilities in response to dynamically changing situations—something that would be familiar to an early cybernetician.

In the real world, prediction rapidly gets hugely complicated. There are many interconnected variables, and small errors in individual judgments compound to produce dramatic predictive failure. Not for nothing is strategy more commonly compared to art than science. That's an observation to reflect on later, when we explore the possibility for using machine intelligence to predict, as an aide to human strategists.

Architectures of warbot intelligence

More capable warbots would need to be smarter than a V1. It wouldn't take long. The Second World War prompted massive technological innovation, nowhere more so than in the fields of electronics and computing. The British codebreaking project at Bletchley Park led the way—most famously via the genius of Alan Turing. This astonishing polymath made several enormous contributions that bear on the story here. Most directly, his creation of mechanical, and then electronic code-breaking machines started the digital computing revolution. Today's warbots run on computers first imagined by Turing.

Before Bletchley, and prior to the war, Turing made his mark in academia by setting out a proof for the essential logical incompleteness of mathematics. That may sound esoteric, but Turing's breakthrough had some practical implications for AI. For one, he showed that maths could never capture the full extent of reality. It didn't matter how many true statements, called axioms, you had, you could still come up with a proposition that could not definitively be proved true just by using those axioms. Another famous mathematician, Kurt Gödel had already offered a proof of this incompleteness, but now Turing stepped in to offer further support. And, as a by-product of the proof, Turing casually invented the modern computer.[12]

In his landmark paper, Turing imagined a computer, which he called a 'logical computing machine', ever after known as the universal Turing machine. This machine, he claimed, could perform *any* calculation (hence 'universal'), by disaggregating them into a series of individual steps. These would be marked on a paper tape and fed through the machine's reader. The computer was entirely theoretical—there never was an infinitely long, one dimensional ticker tape, as he proposed. Perhaps it all sounds a little arcane, but Gödel's finding and Turing's subsequent 'computerised' proof of it offered a salutary rejoinder to anyone dreaming that superintelligent computers could solve all the world's problems. It didn't matter how good the algorithm, or how much data you could pump into the machine; ultimately, reality was still bigger and messier than any machine could calculate. There really is, as far as Gödel and Turing are concerned, a ghost in the machine.

Turing wasn't the only pioneering name in computing, regardless of his outsized contributions. Beginning in the 1930s, Claude Shannon, an American scientist, was fascinated by information flows in communication, especially in telephone systems. He demonstrated that logical propositions, like AND/OR and IF/THEN, and binary numbers could be implemented by switches in a telephone circuit.[13] Switches, binary code and Boolean logic soon became part of the basic architecture of modern digital computing. Another significant figure early in the history of computing was the charismatic American, John von Neumann.[14] His contribution is rather less clear-cut; others can also lay claim to what has become known as the von Neumann architecture for modern computers. By this stage, spurred by the war and the problems of fire control, communications and codebreaking, computing was a large and rapidly expanding field. Still, von Neumann, a brilliant mathematician and consummate committee man, certainly played an outsized part in the story of the computer, advocating a design where the instructions for the computer were separate from the machinery of computing (rather like Turing's logic processor and ticker tape).

Today's distinction between hardware and software goes right back to the origins of the computer. In a way, it foreshadowed modern debates in AI about 'innateness'—how much structure would you have to build into a machine, versus how much it could be expected to acquire on the job. In Turing's example, the 'software' program, inscribed on the tape, did a huge amount of heavy lifting. The hardware itself was pretty basic. But the balance could shift, and the boundary wasn't always clear cut, as with the idea of 'firmware', semi-permanent software that helps run the machinery, which in turn implements the software.

And there was another enduring design feature in the von Neumann architecture—memory, on which to store the program and information that the computer would use in its calculations. In Turing's machine, the tape itself served as the memory. In the real world, creating memory was an engineering challenge, initially solved using cathode tubes appropriated from television screens—discrete 'bits' of information would circulate in the tube, being read off the screen with a foil covering and being looped back into the tube—in other words, being 'held' in memory. This pointed to the clunky, improvised and often unreliable machines of the day. Huge, and prone to malfunction, they seemed more the products of the steam age than the IT era to come. Heath Robinson's eccentric cartoons of complex machinery were popular at the time. These comically depicted complicated and rickety machinery, sometimes powered by steam, or candles, bodged together with ropes and pulleys.[15] Some of Robinson's imaginary machines were secret wartimes weapons. In homage, Bletchley staff wryly christened one of their codebreaking machines after Robinson. Life was imitating art. But while their new machine was certainly erratic and complex, it was a departure from Robinson's industrial age fantasies—in fact, it was the first big step on the way to modern computing, with electronic switches that were far faster than any mechanical alternative.

From computers to AI

The 1940s and 50s were a fertile time for computer science. The new electronic computers weren't particularly powerful, at least by modern standards—with tiny memory and feeble processing power. But the artificial machinery that could support Artificial Intelligence was rapidly coming into being.

And from the outset, there were computer scientists who were interested in precisely this question. The terms were rather woolly—what did it mean to be intelligent? What about thinking, and having a mind? These concepts were far from settled in psychology and biological research, let alone computer science. In fact, the term 'artificial intelligence' didn't emerge until later in the 1950s. But that didn't stop early interest in intelligent machines. And while the motivation of researchers may have been the thrill of scientific discovery or pure philosophical conjecture, their funders had more practical concerns. Most especially, some were interested in national security. There were industrial rationales for computers too—from telecoms to manufacturing. But the link between weapons and computing was there from the outset. The first computers, built in the US and UK, were heavily funded by governments concerned with the deepening Cold War confrontation. It was already clear from Bletchley that computers could help with codebreaking and its obverse, encryption. But what else might they do?

The 'boffins' eagerly set to work designing programs that could play chess, or at least chequers (draughts) against human opponents. They were interested in basic research questions—how did a computer reason? Could you make a machine that 'thought' like a human? This meant exploring the fundamental building blocks of intelligence, like perception and memory. In the spirit of the times, many computer scientists were breezily optimistic about the feasibility of their research project. Perhaps that was understandable; they had after all just invented computers. By contrast, their government paymasters were far more interested in more military-salient AI. There was always a tension between the sort of abstract research that interested many scientists and the Cold War need for applied technology. Still, there was enough of an overlap to programs keep the money flowing—at least initially. The scientists' optimism certainly helped. Famously, at one early conference in 1956, participants, now legendary names in the field, judged that in just a few short weeks they would make 'a significant advance' in using computers to tackle 'problems now reserved for humans'.[16] Heady stuff.

Two approaches to mind

Two cognitive tasks with military utility seemed particularly promising—object recognition and natural language translation. Like codebreaking, and indeed chess, both tasks drew on key attributes of the new computers—pattern recognition and logic. In turn, those two attributes played to the strength of two rival approaches to machine intelligence that emerged early on and have shaped the field ever since. As so often in AI, both approaches were inspired by human minds. Understanding more about them will help us understand more about the possibilities for warbot minds and the ways they have developed to this point. The first approach, which came to dominate the field for decades, was to think of AI as a matter of logic. The second, which dominates today, was inspired by brain cells.

Early research in Artificial Intelligence drew heavily on ideas from other academic disciplines—most notably psychology, neuroscience and philosophy. The latter, in particular, was influential in shaping the logical approach to AI that dominated for many decades, at least until the end of the twentieth century. In philosophy, Ludwig Wittgenstein was a towering figure. Lean and ascetic, Wittgenstein took a similarly austere view on language and perception. 'Whereof one cannot speak, thereof one must be silent,' he sternly declared.[17] If you couldn't put something in words, you couldn't experience it as conscious thought. You certainly couldn't reason with it. The labelling of things wasn't just an abstract representation of reality—it *was* the reality.

Herbert Simon, present at the 1956 conference and soon an iconic figure in AI research, agreed. For Simon, thought amounted to symbolic logic.[18] And if the human mind worked like a machine, then surely the machine ought to be able to approximate it. This powerful idea then fed across into psychology, where the mind-as-computer metaphor became increasingly popular during the 1960s. All this intellectual cross-fertilisation had profound implications in all three disciplines. But did the human mind actually think like that? Linguistic scientist Noam Chomsky certainly thought so. He argued for an innate language ability, hardwired into human brains; a kind of Ur-grammar, common to all of humanity.[19] And in fact there *are* specialised physical structures in the human brain implicated in language creation and processing. But neither the computer scientists nor the cognitive psychologists were particularly interested in anatomy. They were describing *functional* characteristics of intelligence. Symbolic logic was, in the terminology, 'platform neutral'. On the face of it, if humans thought like that, there was no reason machines couldn't as well.

Symbolic logic offered a powerful way of conceptualising intelligence, or at least conscious thinking, where we seem able to hold ideas in mind, and shuffle them about. Bits of information can be recalled from memory, and 'chunked' or combined together in ways that are useful, much like on a von Neumann computer. Some neuroscientists use another computer analogy—referring to the conscious mind as a 'global workspace' within which ideas can be recombined, much like the computer's desktop or clipboard. And yet symbolic logic left an awful lot out of human intelligence—where was emotion, or imagination, or creativity? The absence of these would return to dog AI research in decades to come.

Cavalierly, AI researchers pressed ahead with symbolic logic. Computers were good at it, and even if human thinking was actually more complex, there was enough to suggest that symbolic logic was at least part of what was going on. In the AI paradigm, information was perceived, reasoning was applied, there would be some measure of planning ahead, searching through possible options before deciding what to do. The challenge was to provide the optimal solution—the most efficient, rational and logically consistent way to tackle a problem. Symbolic logic was integral in early efforts to simulate the movement of blocks by a robot arm, working out what sequential moves would reorder them in a new combination most effectively. And it featured in efforts to apply AI to board games. These games, like the simulated environment of the robot arm, were useful because they stripped out much of the noisy, messy confusion of reality, allowing the researchers to home in on the core, foundational activities that many believed underpinned intelligence.

Pioneering robots attempted to navigate in the real world in this way. But the results were rather poor, limited by both hardware constraints and the complicated programming required. A landmark American robot of the late 1960s, Shakey, demonstrated the limits of state-of-the-art AI.[20] A Dalek-like contraption that perceived the world through its onboard tv camera and sensing 'whiskers', Shakey would sit idle for minutes at a time while its computer laboriously ground its way through the various combinations and permutations. And Shakey wasn't even moving about the real world, but a carefully simplified lab version of it, whose surfaces were painted and brightly lit to assist its computer modelling of the environment. If there was too much novelty—something in the wrong place, lighting not quite right—the robot was flummoxed. Researchers would say that its intelligence was 'brittle'—unable to cope with novelty.

Shakey was hugely ambitious, combining distinct research sub-fields in computer vision, language processing and robotics. And like many AI projects of the era, it was funding by the Pentagon, through its Advanced Research Projects Agency (ARPA—a D, for Defense, was added in 1972). The Pentagon was an enthusiastic sponsor of many AI research projects, forging links with centres of excellence at universities around the United States, including Shakey's home department at Stanford, but also MIT and Carnegie Mellon: today these remain leading departments in the field. Alongside ARPA, military funders included the Office for Naval Research, the Army Signal Corps and the Rome Air Defense Center. And not just the Pentagon: the CIA was interested in image recognition and natural language processing, including one unsuccessful attempt at face recognition in the 1960s. Research also involved private enterprise. Famous names like Hughes, Bell Laboratories and IBM, along with more obscure outfits, like Panoramic Research and BBN, sometimes spun out by the university researchers themselves. A fertile ecosystem for AI research emerged, a blend of government money, university brains and private sector entrepreneurialism. Add in the political drive from Cold War Washington, DC, and the growing popular culture fascination with robots and the result was a uniquely potent American blend. It was a military-industrial complex, certainly, but far larger than that too.

Half a century on, Shakey looks rather basic to modern eyes. But its achievements were hugely impressive, especially given the problems that were becoming apparent with mainstream approaches to AI. A paradox was emerging: tasks that humans struggled with—like statistical analysis, or probabilistic reasoning—could be crunched in moments, even by these early computers. But tasks that humans found incredibly easy—moving, grasping, even seeing—were proving altogether beyond even the most advanced machines. Even translation was proving difficult.

The first of several so-called 'AI winters' set in during the 1970s, when the limits of this approach to AI became apparent. Shakey was impressive, but in Vietnam war was raging, and even the Pentagon's deep pockets were not bottomless. DARPA abruptly cut funding from 'basic' research on AI—the sort of projects the superstar team behind Shakey was interested in. Instead, the American military increasingly favoured more applied research that might rapidly be fielded in combat against the Communists. Technological innovation continued, and some of it would have marked implications for warbot design to come. There were, for example, great improvements in precision-guided munitions, using infrared or optical sensors. These were intelligent weapons that could autonomously adjust their behaviour as events unfolded, but they were very far from computers that thought like humans. Communications and surveillance satellites proliferated, and work began on a space-based global positioning system. And there were advances that had no immediate impact on AI, but which would eventually be integral to its development—as with the internet, again funded by ARPA. The Pentagon even had a hand in the development of the computer mouse. A unique hi-tech culture was emerging—a hybrid of federal research contracts, private contractors and university research. In the 1970s, it fed into an extensive transformation of American military activity, in which networked digital computers played the key part. This was the sort of applied research the military urgently needed, to 'offset' the conventional military might of the Communists, not rickety, bewhiskered robots.

In large part, the constraints of hardware were responsible for the meagre results. Digital computers were only two decades old, and severely limited in computational power. Playing a decent game of checkers was one thing, but computers still hadn't beaten highly ranked humans at chess, a game that played to their statistical prowess. And there was a larger problem—a philosophical flaw at the heart of the AI project which proved almost fatal to the symbolic logic approach. In the language of Wittgenstein, just because you couldn't speak of something, it didn't mean it didn't matter. The 'frame problem', as it's known, was all about knowledge representation. How could you capture reality in a simple way that presented all the salient information a computer might need to solve whatever conundrum you had in mind? Related to this was the problem of the 'combinatorial explosion'. Many problems AI researchers worked with involved searching ahead through possible scenarios. Which move should you make in chess, or Go? What would the world look like if you moved this block here? What was the optimal route between two points on a map? The basic difficulty is that the number of possible moves increases rapidly the further into the future you look. You may attempt to find a solution through exhaustive mapping of all possible routes, but in all but the simplest environments, you would exhaust your processing power before getting very far, even with modern computers that are far more powerful than those available at the time. The only solution is to ignore many of the options—but which ones?

The challenge is to squeeze all the possible data from the real world into a format to which the machine can apply its vaunted capacity for logic. Alas, the real world is rarely so obliging, which is one reason for concentrating research on narrowly constrained 'toy universes' like chess. Early efforts at language processing and translation amply demonstrate the difficulties. One popular technique here was to relate concepts to each other in so-called 'semantic webs'—basically topological maps of how objects and ideas are inter-related. But language is rather more fluid and inconsistent than this neat approach suggests. Describing material objects is hard enough—try to define a chair in a rigorous and parsimonious fashion. And that's before you get into immaterial ideas, like beauty. What is a beautiful chair? When it came to AI translation, one of the goals of the CIA, the result might be a clunky word for word transliteration that is at best hard to follow and at worst a dog's breakfast. There's not much scope for idiom and metaphor in transliteration: what's French for 'dog's breakfast'? Chomsky might have been right about a universal grammar, yet teaching it to a computer was beyond the best computer scientists of the day.

That wasn't the end of the symbolic logic story, by any means. As the 1980s began, a revival got underway—this time based on 'expert systems'. The idea here was that niche knowledge about a particular issue would compensate for AI's limitations in the more basic, fundamental intelligence skills, like perception and planning. Humans would, in effect, narrow the problem down to a scale where computers might tackle it using probabilistic and logical reasoning. This 'handcrafted' knowledge would act as a 'heuristic' or cognitive shortcut, for machine intelligence, and thereby—hopefully—save the dominant paradigm.

The approach revived AI research, bringing new funding and new commercial opportunities—including in defence. There was no new underlying philosophy of intelligence—this was still essentially symbolic processing. Still, the addition of richly detailed information made a big difference. After the decade-long AI winter of the 1970s, DARPA's interest in expert systems led to a new round of investment in what it termed the 'Strategic Computing' programme.[21] The name was something of a misnomer—the computers weren't about to take over strategy like WOPR in *War Games*, but to carry out particular military tasks. This time the goal was to have fielded systems, not basic research on abstract phenomena. And there was some success. The most remarkable thing is not that the Pentagon's projects of this era eventually came up short, but that they made so much progress. The strategic computing programme included efforts to build a 'battle management system' for the US Navy, able to track the whereabouts and status of the fleet, and even offer suggested courses of action to commanders. This was a pioneering effort to use AI for something grander than controlling a particular weapon. And even if it was pretty basic, it anticipated the sorts of 'strategic' AI we'll consider later.

Perhaps most impressively, they even managed to design a 'pilot's associate'—featuring a functioning cockpit speech recognition system to interact with combat pilots. That was a remarkable achievement given the space constraints and the noise and turbulence of flight. All that rested on clunky 1980s computers and AI that relied on statistical processing of symbols. Less successful though was an autonomous land vehicle for the army. In tests, it was incredibly slow, couldn't detect obstacles smaller than two feet high, and struggled in anything other than optimum lighting conditions. A viable AI car would be another twenty years in arriving.

There may be more we don't know about, still concealed behind the curtains of national secrecy—there have, for example, long been rumours of *Echelon*, an intelligence system shared by western powers that allowed keyword search through mountains of electronic voice interceptions. This was the sort of needle-in-a-haystack problem that AI excels at—searching for patterns in huge complex data sets.

There were successes elsewhere too. Warbot design continued to improve. At sea, the Aegis combat system was deployed on air defence destroyers. Able to identify, track and target enemy aircraft entirely autonomously, Aegis can perhaps lay claim to being the first thoroughgoing lethal autonomous weapon, able to initiate hostilities without any human involvement. In 1988, however, the limits of the system were dramatically exposed when the USS *Vincennes* mistakenly downed an Iranian airliner, killing all aboard. The machine was operating on manual control, with the order to fire given by a human. But the interaction of machine and human appears to have been critical in precipitating the error. In the moments before launching a missile, the Aegis computers re-assigned the numeric identifier for the airliner to a more distant fighter jet. The ship's captain, told the fighter was descending, mistakenly thought this referred to the airliner, and loosed the missile. It was an inauspicious moment for a system that had been almost two decades in development, and which would go on to be a mainstay of the US Navy for decades to come. Another warbot commissioned in the 1970s that came to maturity in this period was the Tomahawk cruise missile, launched from the sea and able to hug terrain, before arriving precisely on target over a thousand miles distant. Still another was the land-based Patriot missile interceptor, which like the Tomahawk entered public consciousness in the 1990/1 Gulf War. All suggested highly refined autonomy—but not much intelligence.

But DARPA's lumbering effort at an autonomous vehicle pointed to the stark limits of AI in the 1980s. It was rather closer to Shakey than KITT, its fictional counterpart in the *Knight Rider* television series. Precision-guided missiles like Aegis and Tomahawk were polished versions of the proximity fused artillery shell—a very limited sort of intelligence. The architecture for AI was improving all the time, with faster computers and better sensors. The problem though was with the machine's brain. Expert systems brought much needed focus and practical application to AI research. But it was an intellectual *cul-de-sac*, without much to offer in the way of insights for future directions in the field. The approach was neither generalisable nor flexible and it certainly bore little resemblance to a human brain.

Some researchers optimistically thought that expanding the knowledge base of an expert AI would produce qualitatively different intelligence. Feed enough information into a sufficiently powerful machine, and voila: real machine intelligence. Douglas Lenat's 'Cyc' was one wildly ambitious scheme, launched in the mid-1980s which sought to encode great swathes of human knowledge and common sense.[22] Rather than just explicit knowledge—'Paris is the capital of France', the programmers sought to capture implicit, common-sensical axioms: 'you smile when you're happy'. They're still at it, having inputted tens of millions of knowledge statements. It was a statement of faith in the dominant paradigm. And it failed, at least if the goal was to produce human-like intelligence. Cyc has become the mother of all expert systems, but still has difficulties inferring concepts that humans would grasp instinctively. Ultimately it's still completely reliant on those dedicated teams of data handcrafters, rather like the AI version of Wikipedia, to which it is, naturally, hooked up. People don't have ready access to tens of millions of truth statements, and yet they can effortlessly generalise across concepts. A toddler doesn't need to be told that sunlight shines from above, or that a wooden table is solid—she just senses these things and can instinctively and flexibly use that knowledge elsewhere.

Expert systems prompted an AI revival, but as the 1990s unfolded the field was stagnant. Progress rested on the increased power of computers, rather more than any philosophical breakthrough. In the late 1990s, decades after AI optimists had predicted, the world chess champion was finally defeated in matchplay against a computer. Deep Blue, we'll see shortly, made headlines not because it was intelligent in any way comparable to a human, but because its underlying architecture allowed it to search deeper into the game than any previous machine. Computer power triumphed over human ingenuity. A breakthrough of sorts, but not perhaps what Alan Turing had in mind when he prophesised in 1950 that:

> at the end of the century the use of words and general educated opinion will have altered so much that one will be able to speak of machines thinking without expecting to be contradicted.[23]

Throughout the cycles of boom and bust, the optimism and hyperbole that were a feature of AI probably didn't help. Herbert

Simon and Alan Turing were typical of upbeat researchers who had thought that solving intelligence was a feasible challenge. When progress proved altogether more elusive it made the discipline all the more vulnerable to criticism. Why were they wrong? In part perhaps it was the pioneering spirit of the time, as a wholly new field emerged, based on the undoubted technical brilliance of early researchers. In part it may have been because the ostensible goal—a machine that thought a bit like a human—rested on a hazy understanding of human intelligence.

Ultimately, reducing thought and mind to the shuffling of symbols misses a great deal about how human intelligence actually works. For one thing, it misses emotions. In humans and other animals, emotions work as a kind of heuristic, or mental short cut, parsing the mountains of information we receive, and making decisions quickly and efficiently on our behalf. It's not just emotions either. Early in their development humans attain basic 'common sense' building blocks of knowledge, that can be combined in many useful ways. Machines lacked that flexibility, and early attempts to program it for them were laborious, clunky and unconvincing. In fact, they remain so today. Machines might use artificial heuristics to narrow their search through possible options to manageable proportions. But these are not a human sort of heuristic, informed by aeons of evolutionary adaptation. And for expert systems, the heuristic was, in any case, exogenous—provided by humans, not learned by the machine.

Fundamentally, the differences had to do with meaning and understanding. Humans were capable of that, but computers, except for a brittle grasp on some hand-coded categories, as in a semantic web, were not. In recent decades, philosophers of mind have become increasingly interested in the notion of 'embodied cognition', exploring the way in which knowledge for us is generated by and situated within a body. From the myriad noisy signals perceived by receptors in the body, the human mind is able to create meaningful representations and build flexible models of reality, updated in near real time. A logical processing algorithm cannot access that experience of reality—no matter how detailed an explanation of the colour red you give it, it never experiences the feeling of red. If the goal was to build human-like intelligence, that was a formidable hurdle. Even if it wasn't, and machine intelligence were judged on its own terms, there were evident constraints.

Marvin Minsky, another titan of AI research, had been present at the creation—the famous 1956 summer conference. He shared the optimism of the others present, eagerly anticipating that within three decades computers would be thinking along human lines. By 2003, however, Minsky had changed his mind, arguing bleakly that, 'AI has been brain dead since the 1970s.'[24] Jack Schwartz, the man in charge of DARPA's expert system programme in the late 1980s was also sceptical. The Pentagon's revived interest in AI had been spurred by fears that rivals like Japan might be pulling ahead. It rather anticipated modern-day concerns about the rise of China, and its growing expertise in AI. But for all its achievements, there was still a gap between hype and reality. For Schwartz, the fundamental problem remained—computers just weren't particularly intelligent, and he had doubts they would be any time soon. Expert systems in particular required highly structured knowledge—there really wasn't any learning on the job. As he put it, the history of AI research was of 'very limited success in particular areas, followed immediately by failure to reach the broader goals at which these initial successes seem at first to hint'.[25]

Greater computer power would muscle through problems that benefited from it, like searching ahead on the chess board. That soon proved enough to beat even the world champion. But common sense? Forget it. AI might be ok for very limited military applications, especially at the tactical level. Yet, even then, performance was modest, especially compared to the millions of dollars of investment and the expectations of those signing off on the cheques. Anything demanding greater insight and understanding was not remotely feasible. After half a century of research, for the military at least, AI remained a souped-up calculator, not a strategic genius.

As the 1990s began, the Pentagon's expert system programme went into terminal decline. For the second time in its short history, AI reality had not matched the hype. A new AI winter loomed. But all that was about to change.

DEEP WARBOTS

On the highway outside Kuwait City, the smouldering remains of Republican Guard vehicles stood as a sombre testament to the power of computer networked warfare. It was 1991, and Saddam Hussein's battle-hardened army had been utterly routed in a matter of days by an American-led coalition. Fleeing for the border, hundreds of vehicles had been attacked from the air and annihilated, the final blow in a stunning defeat.

The Iraqi army had a deserved reputation for resolve, having fought a long and bloody war against neighbouring Iran for much of the 1980s. It was well-equipped with modern weapons—artillery, armour and aircraft, much of it from Soviet supplies. Against them, the Americans had deployed a massive force, and were understandably nervous about taking on a capable enemy. It was only fifteen years since searing defeat in Vietnam had prompted a thoroughgoing reappraisal of America's approach to war. That conflict stimulated the development of new warfighting technologies, such as precision munitions that could follow laser and televisual guidance, or home in on signals emitted by enemy radar. It had also spurred the development of stealth technologies that could hide aircraft from enemy view—a prototype for the famous F-117 stealth fighter flew in the late 1970s. And it prompted new operational concepts too, notably the need for more effective joint action, with air, land and sea forces working in closer harmony. The goal was to offset a perceived Soviet advantage in conventional weapons—by substituting precision and speed for mass.

The new approach required advanced communications—a secure network that could integrate all the information on the battlefield and distribute it swiftly to where it was needed. As ever in military circles, enthusiasts excitedly deployed new jargon. All the talk was of 'network enabled warfare' that would allow 'information dominance', seamlessly connecting 'sensor and shooter'. It was a computerised 'revolution in military affairs'—another favoured bit of terminology. At its heart were great masses of data—and 'big data' would soon kickstart the new revolution in AI research.

New warbots made their public debut in 1991—the Tomahawk cruise missile, memorably described by one correspondent in Baghdad as flying down a street, and turning left at an intersection; and the JDAM, the joint directed attack munition, a satellite guided bomb. Television news broadcasts showed flickering black and white images of a cross-hair settling over a building or vehicle, before the screen flashed blinding white as the target was destroyed. And lastly, the Patriot interceptor batteries, deployed to Israel to guard against Saddam's threats to attack its cities with his crude 'Scud' missiles, a derivative of the venerable Nazi V2 rockets. The brief combat was so one-sided that almost everyone was surprised, including the Americans, who had anticipated a tougher fight. The lesson was clear—combat veterans with capable legacy equipment were simply no match for the new American high-tech way of war.

The innovations built on America's growing dominance in information technologies, and the strong links between its university research community, the Pentagon, defence contractors, and even the more freewheeling high-tech community in Silicon

Valley.[1] Thirty years earlier, President Eisenhower had soberly warned of a 'military industrial complex' that might test the social contract in America. But its unique blend of finance, research and defence had also ensured America's post-Cold War emergence as the world's sole superpower. And here it was in a new guise, readying America for a new generation of war.

If anyone wanted to emulate that technological revolution, perhaps they would have to draw on similar societal forces. It seemed unlikely. As Saddam capitulated, the Soviet Union imploded. The fall of the Iron Curtain was followed by a period of economic turmoil, thankfully peaceful, in the former Communist bloc. Some commentators made the connection explicitly—trying to keep up with America's defence modernisation had overburdened the already strained Soviet economy. Frankly, the sort of individual enterprise that could see Apple Corp founded in Steve Jobs' garage wasn't the sort of thing a centrally planned economy could pull off. At a time when western children were learning programming in the classroom and playing games on their Atari home computers, the Soviet Union was languishing so badly that even leading weapons laboratories lacked modern IT. America had found the right balance between the iconoclastic brilliance of entrepreneurs like Jobs and Bill Gates, and huge, centralised funding marshalled by DARPA and the CIA.

But for all the hoopla, the new warfighting technology wasn't really big on AI. Cruise missiles and precision bombs made up only a small fraction of the ordnance used in the air campaign. And anyway, how intelligent were all those autonomous weapons—all the guided bombs and cruise missiles? Missiles could change course to adjust their flight towards a target, but target selection itself was done by humans. The route followed by the Tomahawks was programmed by humans. The firing decisions were often taken by humans too. The systems sometimes allowed fully autonomous response, but only because of the need for speed. And the networked systems—the command and control that would achieve 'information dominance' over the hapless Iraqi army—that was just high-end IT.

No matter how polished you got it, AI was undoubtedly running out of puff. DARPA's stalled efforts at building an autonomous vehicle though showed just how limited it all was. Like a cruise missile, it was following a recipe supplied earlier, with hierarchical lists and protocols. Do this, if that, then this. Not only was the AI limited, but it had little capacity to improve on its own. Lumbering along at 5 mph, struggling to make out anything smaller than a large boulder—this was emphatically not the stuff of science fiction magic. It would take a paradigm shift for that to happen.

Back to the future

Flash forward to the springtime in 2004 and the warm desert outside Barstow California. DARPA was back in the hunt for autonomous vehicles. Its 'Grand Challenge' required vehicles to navigate 142 miles of off-road course, entirely without human input, within ten hours. There was a $1million prize, and fifteen competitors vying for it on the starting line. But no winners. The furthest anyone got that first year was 7.5 miles. Undeterred, DARPA relaunched the competition, and scheduled a new race eighteen months later. This time there was a winner: Stanford's 'Stanley', a modified VW Tuareg, led home a field of five finishers, jump-starting the modern race for commercially viable autonomous cars.[2] It was a powerful demonstration of the American approach to AI—once again harnessing universities and hobbyists to the defence complex. In the nearly two decades since DARPA's lumbering Autonomous Land Vehicle, computer hardware had become far more powerful, and sensors more sophisticated. But Stanley's apparent triumph masked some enduring weaknesses. Its underlying philosophies were resolutely old-fashioned—modules for sensory processing fed into planning and decision-making modules. Tellingly, while there were obstacles on the route, there was no other traffic, and no pedestrians. And for anyone thinking about combat vehicles, there was no one shooting at the robot cars.

But in the same year of the first Challenge, the future for warbots whirred into action not far away, and again at Stanford University. Andrew Ng is one of the leading figures in the recent story of AI. And it was his DARPA-funded model helicopter that really signalled a new start for AI. Or, perhaps, less of a *new* start than a revival of old approaches from the early days of AI. Rather than program in rules for the helicopter, Ng and his team used 'reinforcement learning' techniques, so that the computer's 'policy module' learned for itself what to do.[3] A handcrafted 'expert systems' approach to flying would suffer the usual difficulties—gusts of wind and thermals would change the aerodynamics for the helicopter in unpredictable ways, and any pre-programmed expert rules wouldn't be sufficiently responsive. But a machine that learned for itself was a different matter. To start with though, the computer learned by watching as humans controlled the helicopter—a sort of 'supervised learning' where the machine could see what output was required and adjust its inputs to the controls until it had essentially fine-tuned its piloting skills.

It is only a mere fifteen years ago, but the approach that Ng and his team adopted set the scene for some of the remarkable developments in AI that make it such a hot subject today. The personal assistant in your smartphone, and the swarm of drones flying in beautiful formation as part of an aerial display. The medical diagnostic app that outperforms radiologists in spotting tumours. The deep fake algorithm that can create convincing imagery, literally putting words in a politician's mouth. The games computer that can humble a world champion Go player and achieve superhuman performance in arcade games. The algorithms that can learn your shopping preferences, track your physical movements and recognise your face among thousands. And yes, the automated car driving on a street not far from you. All these and more are based on a different AI philosophy from the symbolic logic we saw previously—one sometimes known as connectionism.

While not a new approach, it revives one with very deep roots. Back in the 1950s when AI researchers were looking to human intelligence for inspiration, some sought answers in the biology of the brain, rather than the psychology of the mind. Even then some key parts of the puzzle were understood. After centuries of deliberation and somewhat haphazard research on cadavers and animals, there was an appreciation that human intelligence was centred in the brain and not (as ancients had thought) the heart.[4] Moreover, scientists had established that the brain was formed of myriad cells, organised in densely connected networks. These, it seemed, received information via the nervous system, and communicated with one another via electro-chemical impulses—and not, as had been suggested by earlier theorists, some form of hydraulic pressure in the nerves. Establishing that fact alone had been a labour of decades, and there was still much unknown about the process. Among the key debates then underway was the question of how this network activity gave rise to thinking. Where were thoughts, memories and emotions? In particular, how did all the physical activity of interacting neurons give rise to the subjective experience of *being*—consciousness? Another large challenge was to explain how the brain learned to do things—did particular parts of the brain specialise in certain activities, or was knowledge distributed across the whole network? The answer, it eventually emerged, was both.

Perhaps it is unsurprising that the heady optimism in AI of the day should have embraced the architecture of the brain as a model for intelligence, even with all these unanswered questions. But in truth the direction was two-way, as neuroscientists (the field was itself a new discipline) embraced mathematics, logic and computer science as a way of grasping human biology. In 1943, while Alan Turing and colleagues were labouring away at Bletchley on their designs for an analogue codebreaking computer, the American theorists Warren McCulloch and Walter Pitts produced a mathematical model of the neuron, which they called a 'Threshold Logic Unit'.[5] It bore a superficial relationship to the real thing—and that superficiality epitomised the direction of travel for connectionist AI thereafter: the brain would serve as a loose analogy, but only that, for machine intelligence.

Some of the early researchers on this sort of artificial neuroscience were also interested in the symbolic logical approach to AI. It was a small field in those days, with the air of pioneering excitement. One early system applied aspects of both approaches. Oliver Selfridge's superbly named *Pandemonium* comprised a number of 'cognitive Demons', arranged in hierarchical layers, each specialised in some sort of recognition. On receiving an input from below, each demon would 'shout' to the next layer—the volume of which reflected its confidence that it had seen the feature it was looking for. Each demon was of the symbolic type of AI (if you see this, then shout), but together the system was something new: artificial connectionism. The decision demon at the top would somehow make sense from the pandemonium that ensued. Moreover, Selfridge anticipated another feature of modern AI—the volume of each shout would somehow be determined by its prior performance. Demons that were effective in producing the desired result would have their volume boosted.[6]

At about the same time, another seminal connectionist model, from Frank Rosenblatt, made a similar link to biological neurons, in work funded, not untypically, by the US Air Force and Navy.[7] Rosenblatt called his artificial neurons 'perceptrons', and he built physical versions of them to test his theory. As it matured, connectionism became increasingly distinct from symbolic approaches. But winter was coming. A 1969 critique from two AI superstars caught the mood of the moment.[8] Marvin Minsky and Seymour Papert are often credited with undermining connectionism with their rather technical point about the ability of single layer-perceptron networks to compute a logic function. But larger factors were anyway in play—the hardware of the day was limited, and so too were the mathematics. Particularly vexing was working out how to correct the communication between artificial neurons so that the whole system improved. And anyway, symbolic logic was making regular breakthroughs and attracting big name researchers and money. Scholarship is not after all immune to fashion.

Connectionism redux

Even with symbolic logic dominant, research on connectionist AI continued in the background. Some of today's superstar researchers began academic life toiling away in what was often seen as a relatively unglamorous backwater. Facebook's Yann LeCun spent the late 1980s, working on ConvNets, a neural network specialised in visual tasks. Geoffrey Hinton, another titan of the field today, was also plugging away on neural networks in the 1980s—making important contributions to a vital breakthrough in the maths underpinning some of today's connectionism. In the last decade, though, these relative outsiders have emphatically moved to the mainstream. In the latest twist in the winding history of AI, connectionism has become far the most important avenue for AI research. Ng's helicopter was a dramatic demonstration of what was possible. Less glamorously, LeCun's work on handwriting recognition set the stage. His nets were able to detect features in even the least tidy scrawl, and reliably interpret which letter the author intended. Yoshua Bengio, another superstar researcher, was at work in the 1990s on language processing. All these researchers had the idea of using 'deep' neural networks to produce machine intelligence, each net organised into many layered hierarchies—hence the 'deep'.

The landmark demonstration came a few years after Ng's helicopter demo, and he was again involved. In 2012, his team at Google Brain created a sensation by training a deep network to recognise cats—or at least, to recognise pictures of cats culled from the internet.[9] In an illustration of the shifting centre of gravity in AI research, this was a Google funded project, not from the usual suspects at the Pentagon. In time, that would have significant implications for those seeking to capitalise on the most advanced research for military purposes. Like the AI helicopter watching humans in order to learn the basics, the Google system was 'supervised'—tuning its connections on thousands of labelled cat images before setting out to search for more examples.

The cat detector was the first sign of a torrent of deep learning breakthroughs to come. Timing was everything. Artificial nets were an old technique, but by 2012, conditions were propitious for machine learning. The modern revival of connectionism owes to many factors—one of which is undoubtedly frustration with the grindingly slow progress of alternative approaches. But it took more than frustrated scientists to revive connectionism. Three things made the difference—processing power, big data and clever maths.

Ever since its invention in the 1940s, computer power had grown rapidly, even as costs diminished. That relationship is captured in Moore's Law which predicts a doubling in processing power every two years, and a halving in costs. Cheap computer power was essential for deep learning, because the new connectionism made astonishing demands—both for physical processors, and also for the electricity to run them during training. Happily, in the graphics processing unit (GPU) of modern games consoles, AI researchers found the ideal artificial brain cells for their neural networks: cheap, very powerful and widely available.

The second big development was a sudden abundance of data on which to train the nets. More information was being stored digitally, which could be shared at low cost. The internet, of course, was integral to this. Here was the defence-industrial complex creating a technology, the internet, one of whose spin offs would be more useful AI, that in turn would benefit the defence-industrial complex. It's a symbiotic relationship that would impress any cyberneticist. By 2012, the internet had created a handful of incredibly wealthy technology firms, all specialising in data. Facebook and Google led the field in America. In China, Baidu was emerging as a regional hegemon. The new internet giants had enough money to sponsor basic research of the sort that even the Pentagon had eventually baulked at. And they had money to spend on talent too—hoovering up the new AI stars from their university posts: LeCun to Facebook, Hinton to Google, Ng to Google, then Baidu. The new tech giants saw the huge value in using intelligent machines to make sense of the unending torrent of internet data—to find out more about their customers and offer new services. Like the military before them, they shared an interest in pattern recognition, search, and image and language processing.

When it came, the revolution would be feline. But that was only the start. It's barely a decade since Google's cat breakthrough, and the pace of innovation is still accelerating, another important feature of the modern AI landscape. Now deep nets can recognise individual faces, lip read video imagery and produce a simultaneous transcription. They can anticipate what's in an incomplete image—for example creating a predicted face on the basis of many thousands they've seen before. They can even create wholly convincing photographs of people who've never actually existed.

Artificial neuroscience

The third factor in the deep learning revival was more arcane, but critical. This was the increasing technical sophistication of the neural networks that underpinned connectionism. One important development was the discovery of 'backprop', or backward propagation. This was a key bit of maths that allowed the artificial neurons in the connectionist AI to learn effectively. With multiple layers in the modern 'deep learning network', and with many more neurons and connections between them, working out the optimum connections between them had been fiendishly difficult. That's where backprop comes in.

Neural networks are sometimes trained in a supervised manner—learning, like the cat detector, by looking at labelled training data. But supervised learning wasn't the only possibility. Labelling all that data for supervised learning could be a laborious process—akin to the handcrafting of information needed in an old-fashioned expert system. On the internet, much of that work was done for you by users—'this is a cat'. More efficient would be a net that could train itself on data—picking up whatever information it learned to be useful in improving its performance. That sort of unsupervised learning is a specialism for Bengio, and a growth area in AI today. Advocates think it will produce a more flexible sort of AI that can learn like a human does, without millions of examples. A third approach was the one that Ng had used—reinforcement learning—where the machine interacted dynamically with its environment and tuned up its connections on the basis of some reward. That reinforcement learning approach to deep learning has underpinned many of the landmark breakthroughs of recent years, including some that have direct relevance for warbot designers. Like the researchers, they are looking for AI that can learn dynamically, through seeking to optimise its performance. The use of electronic games as a test-bed for reinforcement learning has been a particular research focus. The attraction to military minds is obvious—games are adversarial, and the goal is to win. The differences, however, are also profound, as we'll see.

2015 saw the public arrival of DeepMind, a relative British newcomer to AI research, newly acquired by Google. DeepMind's founder Demis Hassabis had trained in neuroscience, and he was explicit: DeepMind intended to create 'general' AI, with the attributes of human intelligence. Its first landmark breakthrough was an eighties throwback: classic Atari arcade games. The scoreboard in Space Invaders is an ideal motivator for reinforcement learning. Like dopamine in the brain of a teenage arcade goer, the network responded to the reward of a higher score—pruning its connections accordingly.[10] Combine that with a ConvNet that would capture what was happening on the screen, and the AI was all set to play a mean pinball, or rather Space Invaders. In fact, DeepMind's breakthrough arcade AI playing Atari performed creditably on six games, surpassing expert human level on three. Six years later, its latest version, Agent57, now performs better than humans on all 57 Atari 2600 games.

DeepMind again illustrated the new landscape of AI research—a hitherto obscure company, rapidly acquired by Google, which proceeded thereafter to draw in research talent, creating a snowball effect that continues today. This was civilian research, and abstract rather than applied. Space Invaders battled in a very simple virtual world, but the real-world possibilities eluded no one with a military mindset. DeepMind amply demonstrated, like Ng's helicopter, the potential for superhuman skill in physical control, and also the particular strengths of ANN in optimising scores. In the Atari games, DeepMind's algorithm knew nothing about its world except the score—but that was still enough to produce novel and highly effective tactics—in one game, Breakout, sending its bullet through a narrow channel to ricochet destructively behind the banks of approaching coloured bricks.

And there was evidence here too of something else symbolic logic AI lacked. The new algorithm became successful across all manner of arcade games: DeepMind had produced a flexible intelligence. Or more flexible at least than earlier AI which would fall over at the slightest logical inconsistency, novel circumstance or coding error. Flexible and robust: more military virtues. But there was another, less welcome feature—the internal workings of neural nets were a mystery, even to those who created them. There was no way of knowing what each individual neuron was contributing to the whole. Knowledge was 'distributed' across the entire network. That mattered for pure research, but even more for the practical application of neural nets to sensitive fields, like medicine, transportation and war.

The American military was watching Google and DeepMind's stunning achievements. Deep learning held clear promise for tasks they'd long been interested in. From autonomous warbots to intelligence analysis, the allure of connectionism was irresistible. And there was further incentive, from the long running wars in the Middle East, in Iraq, Afghanistan, Syria and elsewhere. Hundreds of Americans were being killed in these 'forever wars'—fighting elusive and deadly enemies. Perhaps autonomy could offset that. Defence contractors were set to work competing for crewless platforms—fighter aircraft, submarines and boats, even land robots. And of course, for the increasingly ubiquitous drone, synonymous with modern warfare. These wars, fought using the same high-tech information-heavy concepts that had twice routed Saddam Hussein's army, were producing a deluge of information, including from satellite and drone surveillance, telephone and electronic intercepts.[11] There were increasingly numerous sensors on the battlefield, and limited capacity to analyse it.

The Pentagon was still funding AI research, including via its traditional DARPA model. But now there was a new challenge—how to make use of the new technology landscape, with Google, Amazon and Facebook sponsoring cutting-edge research. In 2017, the Pentagon stood up an 'algorithmic warfare cross functional team', known as Project Maven. The team would consolidate 'all initiatives that develop, employ, or field artificial intelligence, automation, machine learning, deep learning, and computer vision algorithms'.[12] It was a small team, initially, but would grow rapidly. And one of its main contractors on image recognition? Google, of course. The stage was set for the arrival of deep warbots.

Hype or hope?

The deep learning revolution has produced a new wave of AI hype, only some of which is justified. In rapid succession, connectionist AI delivered results that would have amazed earlier students of symbolic logic. At DeepMind, Space Invaders gave way to Go, a hugely popular board game in the Far East, and a 'toy universe' too vast to master by tree search alone. Then came multiplayer strategy computer games, like StarCraft II. Elsewhere too, there were incredible demonstrations of connectionism's potential. ConvNets proved that they could recognise more than handwriting and cats—outperforming clinicians in a range of medical diagnostic tasks. Algorithms refined the control problem, producing mesmerising swarming drones, like those above Pyeongchang in Korea that formed a glowing set of Olympic rings ahead of the 2018 winter games. The iPhone in my pocket produced convincing real-time translations of foreign road signs and menus, at the point of a camera.

The achievements of researchers working on connectionism were certainly impressive. Machine learning systems could manifestly produce purposeful action in a wide variety of complex, messy environments. Their behaviours could surprise humans by their novelty and—within certain narrow bounds at least—surpass human performance. And all this could be rapidly shared—once an algorithm was trained, it was a trivial matter to upload it elsewhere. These cloned algorithms accelerate the learning process dramatically. For warbots designers, the implications are profound: imagine the individual robots in a swarm observing what works best, tuning their own nets accordingly, and sharing the information with other robots.

But there were significant limits to the new approach. This was a distinctly non-human intelligence. It lacked our ability to understand, to display what we might call 'common sense', an ability to relate objects and concepts, and to intuit deeper meanings behind its observations. Logical AI could do that, albeit not very convincingly, with its semantic webs. But connectionism wasn't even at the races. And while the same algorithms were flexible enough to be applied to any sort of reward-optimising task (like the score in Space Invaders), they were flexible in a narrow sense only. The human brain is hugely parallel, with multiple systems working away on various tasks. Sometimes the same neurons are even involved in multiple systems. We're able to make connections between systems that are specialised in different tasks. Take reading, which for us integrates systems that evolved to do something entirely different.[13] Reading piggybacks on visual systems designed to detect contrast and edges, like LeCun's handwriting ConvNets. But it also connects to higher order processes that attach meaning to the squiggles, to memories, to systems specialised in speech production and understanding. There's much still to learn about how the brain does this integrative wizardry. In real time, we create a coherent internal reality that feels rich and whole. It's clearly not just about training a network so that it can sense raw data. There's a top-down element here—the mind creating an expectation of the world which gives order to the deluge of incoming data. Sensory information arriving at the human brain is not interpreted passively, but rather according to some understanding in the brain of what information *should* be arriving. The brain models reality, making predictions. In short, the brain is hugely intricate with some powerful emergent properties that today's AI wholly lacks.

Moreover, there are also large differences in the *purpose* of intelligence. Our cognition might be driven by some ultimate purpose (to survive and reproduce), but often it's not clear exactly what goal we are optimising. How does the existential joy of riding a racing bike down a country lane on a summer evening relate to the narrow goal of getting the best score in a 1980s video game? For that matter, how do we arbitrate between all the many goals we might be pursuing at any one moment? In particular, how do I prioritise a goal that demands cognitive resources now, but delivers no obvious payoff? Writing this book, for example, requires hard work over an extended period, with no scoreboard ticking over. I could be out with friends instead or doing a thousand other things that would be more immediately rewarding. DeepMind's Atari conquering AI was relentlessly focused on the scoreboard. But it struggled at one game in particular, Montezuma's Revenge, precisely because there was no immediate payoff. For humans, patience, or 'delayed gratification', is a useful trait. So too are traits like curiosity or playfulness—a willingness to experiment, sometimes with no particular goal in mind and no immediate sign of reward. These attributes play important roles in human development, as children learn to apply the basic conceptual building blocks of cognition to a wide variety of challenges. And they're important too when it comes to war—which is about much more than maximising the score on a console game. In war, we decide what we want and how hard we'll fight for it.

There's no ready analogy for these higher-order processes in machine learning. Artificial neural networks are highly specialised, efficiently computing optima to narrow goals. If you want to spot a tank in the undergrowth from an aerial drone, that's probably enough. So too if you want to fly a helicopter—that sort of complex control problem is child's play to them. Actual child's play, however, is a different matter. Right now, connectionist AI is much closer to a souped-up calculator than it is to the stuff of 'mind' and 'thought'. The connectionist revolution certainly hasn't reached a dead end, as Minsky thought of symbolic logic. More computer power and more imaginative designs for algorithms are inevitable. But as a paradigm, there do seem to be limits to the sort of intelligence it creates. If the goal of AI is to produce a richer intelligence—perhaps a machine that seems a bit more human, it's not immediately obvious that connectionism can deliver.

Still, no one *knows* for sure whether it can. On one, rather optimistic, view, the brain is just a network of networks: Yes, it's far more complicated than even the largest neural net, but for some enthusiasts, that's just a matter of scale and engineering complexity. The emergent properties of the human brain, perhaps even consciousness itself, will also appear in artificial nets if only they're made sufficiently complex. Yoshua Bengio, seminal in the revival of connectionism, is in this camp. He believes that neural nets can represent many of the higher order concepts and relationships that brains do. Ideas in the human brain are, after all, just patterns of neurons. Make nets sufficiently deep, and construct networks of networks, and even the sort of abstract reasoning that humans do becomes feasible.[14]

On the other hand, sceptics like Gary Marcus argue that connectionism misses something fundamental about the human mind.[15] A bigger, more densely connected neural network won't cut it. It's not clear exactly *what* that ineffable quality is, just that it somehow adds something to the narrow, goal-oriented attributes of the neural net.

For those who agree with Marcus, solving this problem needs something new. It might, for example, require a hybrid combination of logical processes with the input-output optimisation of the neural network. That would somehow blend the structural and functional approaches that characterise the human brain. Doing so might imbue the hybrid machine intelligence with some of the conceptual building blocks of intelligence that allow children effortlessly to adapt to novelty. The hybrid machine might, like children, learn new skills on the basis of limited exposure to examples, rather than many thousands of training examples, as with the cat database.

But even so that would still miss something about our intelligence. Our capacity to integrate multiple complex processes in pursuit of complex goals, say. Or our ability to conceive different possible futures, to gauge what other minds are thinking. Or our ability to be creative, or emotional, or instinctive, or any of the many other things that make us distinctively human. Achieving that would be to create a formidable intelligence, one that would bring us closer to the Hollywood vision of AI, and a very long way indeed from Grand Challenges and Shakey robots.

Master tactician, moron strategist?

All that philosophy of artificial minds matters deeply for those thinking about AI's role in war, especially now that AI is poised to play a significant role in combat. AI research has already made important contributions to defence—but in truth, these were severely limited by the technology. Guided missiles were possible, as the Tomahawk assault on Baghdad demonstrated, but controlling an aircraft in a dogfight was another matter entirely. Until Ng, it looked like science fiction. Translating foreign language material hoovered up by intelligence systems was possible. In fact, one of America's most dramatic technological breakthroughs of the Cold War, the development of the stealthy F-117 Nighthawk, owed something to just such an algorithm. An obscure Soviet physicist had discovered an interesting property of radar waves shedding off flat planes. There being no obvious national security use for the finding, the incautious Soviets published his paper in a Russian language technical journal, one read by the machines of the National Security Agency.[16] But machine translation using traditional symbolic AI usually made for extremely clunky transliterations that were sometimes barely comprehensible.

Radical improvements to these problems arrived in the twenty first century with the reboot of connectionism. The modern AI revolution is still only a few years old, and it continues to amaze. There are 'deep fake' videos, with Barack Obama realistically speaking words of an impersonator. AI can write convincing financial and sports reports, with the data heavy prose playing to its strengths. It can enhance the resolution of blurred images, and identify faces caught on CCTV cameras. Intelligent machines are at work designing new drugs and new composite materials. There are reusable space rockets that land on a sixpence. Clearly when the task at hand suits the strengths of connectionism, the results can be breath-taking. Give a machine a mountain of data, and some narrow goal to optimise and it's obvious that computers are not just intelligent, they are super-intelligent.

But the gulf between AI and human intelligence remains profound. And it maps neatly onto the difference between tactics and strategy in war. The subject of endless War College debates about what those terms mean, one useful distinction is that tactics are focused on to the immediacy of battle—a violent physical contest between opposing armies. Strategy takes the larger view, both in time, and space. Planning certainly matters in tactics, as does reflecting on what your enemy may do, but there's a reductive immediacy to battle. It's about what you can do to prevail in the moment. It's also an arena where training matters—of techniques that have been methodically learned and practiced, and that have been proved in other battles. Fire and move, fire and move. That sounds a little like the repetitious supervised training of a neural network: cat, cat, cat, cat. Now, what's this? A cat. Mindless, but effective.

Connectionism is perfect for tactics. It will sense things on the battlefield and identify them. Pattern recognition is its forte. It will have a perfect recall of the steps that have worked previously, and it can share its learning instantly with cloned systems. Tactics is about delivering firepower at the right time and place—and connectionist AI is brilliant at this. It can manoeuvre with increasing acuity and it's nothing if not precise. Perhaps its understanding of context is imperfect, and its internal processes are a black box. No one knows what steps each individual neuron is contributing, and the machine can't tell you what's going on inside—it just does its optimising thing. But connectionism's inherent advantages handsomely offset these comparative weaknesses. And anyway, enthusiasts argue, there will still be human combatants on hand to compensate. Militaries are keen to stress the continued presence of a 'human in the loop'—making the ultimate decision about violence.[17]

Strategy on the other hand is far more demanding. It is less bounded and less mechanistic. There's no recipe book for success that you can train your neural net on. The cognitive challenge of war is far more complex than the cognitive challenge of battle. Strategy requires things neural nets aren't good at, like imagination, creativity or intuition. Object recognition, natural language processing, robotics—none of these are particularly useful for the strategist. If anything, strategy plays more to the strengths of symbolic logic machines—if I do this, then they do that. Except that even there, the challenges are formidable. There are very many possible steps that each side can take at any moment—and the action isn't iterative, like chess, it's dynamic: Clausewitz likened it to a wrestling match. And while he also argued that war had its own 'grammar', it's certainly not a grammar that any symbolic logic machine can capture.

A Cold War example illustrates the difficulty neatly. By the 1980s, the superpower confrontation was as grim as at any time since Stalin's day. Longstanding KGB chief Yuri Andropov had become increasingly paranoid about the possibility of a surprise American attack that might at a stroke wipe out the USSR's ability to strike back—an 'assured second strike capability' in the jargon of deterrence theory. It didn't help that the Americans were deploying Pershing II medium-range missiles in Europe that might allow just that.

AI apparently played a part in this bleak Soviet assessment. Early in 1981, Andropov's KGB had run an analysis of the Cold War situation on a new computer system. The British spy Oleg Gordievsky reported to his handlers that the new Soviet computer modelled the correlation of forces, including material and psychological factors on either side. A recently declassified American intelligence report, written in 1990, confirms the details. The CIA assessed that the Soviet system, called VRYAN (meaning Sudden Nuclear Missile Attack) was 'used as a tool to predict US strategic intent'.[18] The computer's shocking conclusion was that the balance of power was shifting decisively in favour of the US. Materially, this was certainly true—but material capabilities and intentions weren't the same thing.

It's still not clear what the capabilities of this mysterious computer were—the Soviets were way behind America in computer technology and AI research. Even the best American computer would have been wholly unable to render a convincing judgment about the intentions of rivals. The CIA reported that VRYAN went online in 1979, the era of Shakey and Pong. Gordievsky's intelligence suggested that the computer employed a database of some 40,000 weighted elements, apparently based on the sort of political, economic and military factors that the KGB thought had proved decisive in the Second World War. Several hundred KGB analysts input the data, which came from across the Soviet government; and regular reports on the balance of power were sent to the Politburo.

By 1983, Andropov, now leading the Soviet Union, had developed an acute fear that the Americans might make a sudden military move. For his part, President Reagan certainly bristled with anti-Communist sentiment, but he had no intention of starting a nuclear war. Andropov compounded the computer's error, instructing all KGB stations abroad to find evidence of an impending American strike. Finding evidence, that is, to validate VRYAN's analysis. Meanwhile, and unaware of all this, the NATO allies launched into a large-scale wargame, Able Archer, intended to test their response to an international crisis; but not, emphatically, to cause one. Debate continues about the degree to which the exercise actually brought war closer—the modern consensus is that the Soviets were genuinely alarmed and fearful that hostilities were approaching.

As the American intelligence analysts noted, it was never clear how far VRYAN influenced Andropov's thinking. But, they added, if the computer was at all accurate, it would certainly have shown that the Soviets were growing vulnerable to a disarming first strike. And there was another stunning revelation in their report—the Americans had secretly built a computer model to emulate VRYAN, which attested to its bleak findings.

It might, the authors suggested, seem 'absurd' to think of political leaders being swayed by a computer model, but there were some compelling reasons to think they could be. First, many of the Soviet leaders were formally trained as engineers, and the computer's formal modelling played to their scientific bent. Secondly, VRYAN was pushing at an open door—the leadership, especially Andropov, was already deeply pessimistic about the deteriorating strategic situation, and so primed for bad news. Finally, Soviet military doctrine, the analysts noted, stressed the need for massive nuclear pre-emption in times of conflict. In any confrontation, it would be vitally important to land the first nuclear blow.

The Soviet computer thus showed the dangers and limitations of using AI for strategy. By no means all of the flaws lay with the machine intelligence. Andropov was guilty of confirmation bias—finding information that fitted his pre-conceived ideas—and of mirroring—assuming that when it came to pre-emption, his American enemies saw things the same way he did.

Perhaps AI enthusiasts in national security should stick to what it's becoming expert at—tactical warbots. Over the next two chapters we'll explore the implications of those in battle. And we'll see why it's unlikely that warbot designers will stick to that—no matter how cautionary the tale of VRYAN.

i-BATTLE IS JOINED

Somewhere in the remote Syrian-Iraqi borderlands, about five years from now, an American Special Forces unit approaches an enemy compound in the back streets of a dusty town. This is war in the 'grey zone'—a messy battlefield, with many actors and complex, shifting allegiances. It's a war that's been going on for years already, and it's likely to be the place where the new AI technologies we've been talking about are first used in combat. What can we learn about how AI will change warfare from thinking about near-future combat in scenarios like this?

We'll return to the A-Team later. First, a note of caution. Experts sometimes refer to 'technological determinism', the alluring idea that technologies will lead to dramatic social changes on their own—in our case to changes in warfare. This myopic view on tech overlooks two things—first, the way in which technology is itself a product of wider societal processes. These shape the creation of the technology itself, and also how it is employed. And second, such determinism overlooks the continuities in warfare that persist despite the introduction of a technology. While it is tempting to see an exciting new technology as the harbinger of radical change, there remain some deep continuities in warfare.

Technology, in short, isn't just 'out there', emerging like a bolt from the blue, and unleashing transformative change. It is an artefact of culture. In this case, AI emerged from a society with a tradition of scientific innovation, entrepreneurial capitalism and a large national security establishment. The US welcomed skilled immigrants and offered a way of life that readily attracted them. This vibrancy that has been difficult for other states to emulate, even in the developed western democracies. An alternative culture like China's has a very different model for defence industrial activity and scientific research. There is plenty of entrepreneurial capitalism there, but also far more extensive state control, fewer personal freedoms, and greater corruption. The jury remains out on whether that culture can produce intellectual and commercial breakthroughs in high technology like those in the US. But whatever does emerge will be distinctively a product of that culture, just as it is in the US, Israel, the UK and elsewhere.

Scholars also distinguish between the nature and character of war—arguing that the character of war changes radically between cultures and through time, but that there is some essence that persists everywhere. This useful distinction has allowed theorists of war to reach for general points that might be sensitively applied to war across cultures. That's what the Napoleonic-era theorist Antoine de Jomini was driving at with his idea that war could be reduced to some fundamental principles.[1] Plenty of other theorists agreed with that, though there was considerable variation about what they actually were. 'Principles of war' of one sort or another are still in use today, with variations, by many militaries. The British version stresses importance of surprise and of concentrating force at a decisive point. The need to maintain morale is also highlighted, as is the need for unified effort. These, advocates argue, are timeless features of good strategy. But problems arise if they are taken rigidly as a template for what needs doing—then tactics and strategy become akin to a mechanistic list for commanders. Too dogmatic an adherence to lists and war starts to look like the instruction manual for a symbolic AI warbot—do this, then do that.

Clausewitz, Jomini's near contemporary and rival theorist, was sceptical about this sort of thinking. He eschewed the notion that war could be reduced to a mechanistic framework. That might do for the more prosaic intellect, but not for the best commanders. Clausewitz also identified some general features in war too—but his were far from rigid. Instead he rooted them in an understanding of human psychology—ensuring their timeless relevance. The nature of war, he thought, was ineffably human. This meant it required good judgment. His ideal commander was a 'genius', by which he meant someone adept at making decisions amidst uncertainty. He understood that war involved plenty of emotion and chance. And though he put much emphasis on the material dimension of fighting—which side had the best equipment, or the most men—he also stressed that morale was critically important in determining victory. Today, it's Clausewitz, not Jomini, who is studied in War Colleges around the world.

And when it comes to the impact of AI, that Clausewitzian distinction between nature and character is useful. My claim is that AI alters the nature of war by introducing non-human decision-making, which would be an unprecedented development.[2] 'Military revolution' is an overused and imprecise term. What counts? Composite bows? Professional armies? Perhaps the discovery of gunpowder or even nuclear weapons? I would set the bar rather higher than a new weapon or concept—reserving the label for something that changes the very nature of warfare. For me, like Clausewitz, that nature originates in war's psychological essence, making AI revolutionary. Frank Hoffman, by contrast, acknowledges the radical, perhaps even revolutionary, impact of AI on warfare, but insists that the essence of conflict—'politically directed violence fraught with friction'—will endure.[3]

While we have undeniably emphasised different aspects of war's 'nature', our views are entirely compatible. Like Hoffman, I think war will still be violent and motivated by politics, even with intelligent machines involved. After all, politics merely implies that groups of humans have clashing goals. And, like Hoffman, I don't foresee the human element of war disappearing any time soon. Even if machines make some, or many decisions about violence, war will remain something that is done by, and to, humans. Unlike an earlier generation of technology evangelists writing after the 1990/1 Gulf War, I don't expect for a moment that even the most sophisticated AI will reduce war to something predictable, even computable. There'll be no lifting the 'fog of war' with even the most powerful artificial superintelligence, something we'll explore further in later chapters. Still, the advent of warbots means that decision-making in war will no longer be entirely, or even (eventually, perhaps) mostly human.

The takeout for our Special Forces detachment—their smart weapons and autonomous systems are the products of a particular culture. This shapes their technical attributes, but also the ways in which they are employed. And, powerful though these autonomous systems may be, the approaching battle remains indelibly human.

Fighting power

Clausewitz's further reflections on materiel and willpower are a good way of thinking about the impact of AI weapons in battle. If we focus too much on the machinery of war, we miss that richer terrain, where human attributes shape the outcome of war. Technology grabs attention—it's why thousands of people attend air shows and assemble by the harbour breakwater in Portsmouth to watch the Royal Navy's fleet come and go. But battle is about more than technology.

Consider an example: If, during the Great War, defensive technologies, like the machine gun and the trench system were so powerful, then how did the big breakthroughs occur in the last year of fighting? After years of stalemate, and futile mass assaults on well-defended trenches, with interlocking fields of fire, manoeuvre made a dramatic reappearance on the Western Front in 1918. First, the Germans broke through in their Spring offensive, and when that petered out, the Allies counter-attacked, bringing the capitulation of their enemy. Perhaps it was the introduction of the tank—a new shock weapon that could power through barbed wire entanglements, cross the crater filled no-man's land and storm enemy positions. Perhaps not. The tank was certainly an effective weapon, and later became a central part of modern land warfare, around which land armies were structured. But earlier British offensives with tanks had failed to achieve a decisive breach. And the Germans made their breakthrough in the Spring of 1918 without its assistance—they had barely any. Instead, they used a new infantry concept—advancing in small groups of 'stormtroopers', seeking cover in dead ground and infiltrating enemy lines, rather than attacking in a mass wave, as previously. The effect was devastating, as German forces poured through Allied lines and advanced miles beyond. After years of static, attritional warfare, suddenly movement was restored to warfare.[4] Clearly technology plays a part in explaining battlefield performance—but equally clearly there's more to it than that.

In its modern doctrine, the British Army talks about 'fighting power'—a rough measure of how effective a given force will be in combat.[5] There are three elements to this useful schema—the physical, conceptual and moral. Technology most obviously affects the physical, changing the capabilities at the disposal of commanders. And technology itself may even suggest suitable concepts for employing that technology battle—as the cliché has it, if you have a hammer, your problems look like nails. But not always. The same weapon systems can be employed in radically different ways by rival armed forces, reflecting myriad cultural factors. The third element of fighting power, the moral component, is sometimes overlooked when it comes to discussions of technology and war. That's unfortunate, because the moral element connects the technical and the human—reminding us once again of war's deep human essence, no matter how important machines become.

We can see all three elements of fighting power combining in battle after battle. The 1940 Battle of Britain, a classic example, can help contextualise our assessment of battle with warbots.[6] That summer, as the Luftwaffe and RAF clashed in the skies over southern England, technology was clearly an important factor shaping the outcome. In performance and numbers, the German and British aircraft were closely matched—the Me109 and Spitfire, the premier air superiority fighter aircraft for each side, enjoyed similar rates of climb, ceilings, manoeuvrability and armament. But in its recently developed radar technology, the British had one large advantage; they could estimate when a Luftwaffe attack was coming, and position aircraft to stage an interception. To do that, they needed to work out a concept of operations, and this conceptual approach is arguably what turned the battle decisively in their favour. Shortly before the war, the RAF developed an integrated air defence system, in which radar tracks, combined with human observations of enemy forces, would be fed into a command and control hub. This HQ coordinated the deployment of scarce fighter forces, marshalling them into position to intercept the enemy. The final element in the concept was the tactics employed by the fighters. First, where possible, the more capable Spitfire would engage enemy fighters, leaving the Hurricanes to pursue the slower, lumbering German bombers. Second, the fighters would operate in small batches of four aircraft, approaching initially in a loose line abreast, ideally from above.

In contrast the Germans had no viable concept of operations. They flicked uncertainly from one target type to another—seeking to destroy enemy airfields, repair factories, and even (albeit in rather desultory fashion) radar stations. Finally, they launched the Blitz, a bombing campaign against London, an attempt to damage industry and shatter civilian resolve for the fight. There was considerable damage and disruption, but the effect was to redouble British resolve, and prompt devastating counter-attacks against German cities. Meanwhile, in the skies over Kent, battle raged. It was a close-run affair, despite the inherent British advantages. Damaged RAF aircraft might be recovered, and reserviced, before returning to action. Downed airmen could likewise return to their units. The Germans had no such advantage. And the Luftwaffe was worn down by the tempo of operations, during the battle and in the preceding *blitzkrieg* advance across western Europe. Pilots were exhausted, equipment degraded. The moral component of fighting power weighed in the overall balance. Little by little, imperceptibly on some days when both sides suffered dozens of losses, the balance of fighting power shifted in the direction of the RAF. At last, Hitler sounded the retreat, to refit his forces and strike out eastwards into the Soviet Union. The Battle of Britain provides an excellent illustration about the importance of technology on battle. Radar was a radical new invention and gave the British a telling advantage. But the technology didn't stand alone. It was embedded within a concept of operations, and it was that combination, allied to the motivation and relative freshness of the British airmen and their marginal advantage in terrain, that brought eventual success.

So when we are considering our AI equipped group of special forces soldiers in Syria, we must keep in mind the allure of technological determinism. A technology might seem otherworldly, and it might offer dramatic new capabilities, far beyond those of the enemy, but that doesn't mean we can reach easy conclusions about who would win any confrontation. If war was that readily determined, we'd simply need to input all the relevant information into a computer, which would spit out the victor's identity, and the belligerents need not actually fight at all. One of the defining features of war is uncertainty—and that can be uncertainty about fighting power. Who has the best warbot, which concept of operations is best, and—crucially—how hard will each side fight? That human element of war, so central to Clausewitz's thinking, is unlikely to go away, even if AI makes it less central to the actual business of combat.

Tactics in i-Battle

What can we learn by combining our understanding of strategic theory, from Clausewitz, Jomini and elsewhere, with our understanding of modern AI capabilities? Our starting point should be that combat is a very human affair, and likely to remain so for some time yet. Theorists today refer to 'human-machine teaming' to describe the interaction of AI platforms and soldiers. For the foreseeable future this will be the reality of combat. Visions of all-robot armies belong to the far future—especially when it comes to ground combat. With autonomous driving demonstrated, at least in principle, DARPA's next 'Grand Challenge' required competitors to field an autonomous robot able to operate in a cluttered and hazardous environment. The agency may have been thinking about recent calamity at the Fukushima nuclear plant in Japan, where the risks from radioactivity made human presence highly risky. But those are precisely the skills needed for operating in ground close combat too. The robots failed to impress.

We've become familiar over the last decade with seeing the increasing dexterity of robots produced by the firm Boston Dynamics. They supplied the ATLAS robots used by competitors in the Grand Challenge. ATLAS is an impressive humanoid machine, capable of bounding from one surface to another.[7] It's a small imaginative leap from seeing that to visions of Arnie's *Terminator*—a relentless killing machine, that as one character says, 'absolutely will not stop, ever, until you are dead'. Thankfully the comic failures of the rescue droids in DARPA's competition suggest Judgment Day lies some way distant. Opening a door handle was too much for one machine, which overbalanced, toppled and was unable to regain its feet.[8] Definitely more Shakey than Arnie. The problem is that Boston Dynamics' machines *look* amazing, and indeed are impressive. But they are remote controlled. AI controlled gyroscopes keep them upright and agile, but not much more. Their capacity for autonomous decision-making is, at best, undemonstrated in these corporate video releases. And if the state of the art in the Grand Challenge is anything to go by, it won't be that great.

Ground combat is particularly hard for AI not just because of awkward terrain, but because it involves other humans. I'll save a fuller discussion of the human dimension until the next chapter, when we consider the connection between autonomous cars and strategy, but for now it's worth flagging. The major problem with autonomy for car manufacturers isn't navigating the physical landscape but knowing how the humans in it will interact with your machine. In autonomous cars that's an ethical issue. It is too in combat—a warbot must understand who is a combatant and who is on which side. But it's also essential for fighting power. The machine needs to gauge how much effort is needed to win any engagement; for now it lacks that contextual understanding.

Some aspects of combat are more susceptible to automation. Ground combat is particularly challenging, but offshore and in the air the terrain is simpler. It's easier to manoeuvre and there are fewer humans whose intentions must be weighed. This is a longstanding trend—we've already encountered cruise missiles and acoustic torpedoes far back in the history of automated weapons. And in the modern era these have been joined by uninhabited aerial vehicles, UAVs, including the famous Predator drone, iconic and controversial weapon of the long-running American wars in Iraq and Afghanistan. Reusable drones go back further than that. My personal favourite is the D-21, an American reconnaissance drone from the height of the Cold War. Looking like a miniaturised SR-71 Blackbird, the sleek and sinister D-21 was designed for long range, high altitude and high-speed spying missions, and flew four over China beginning in 1969. Tactical surveillance drones were in use by the mid-1970s, but the technology really came into its own with the deployment in 1990 of the US Global Hawk. Then as now, most of these drones were employed for surveillance, not combat. But then early in the present century, the Predator was equipped with Hellfire missiles to solve an increasingly pressing problem—the lag between identifying a target and getting a weapon system in range quick enough to kill it. An armed drone made attacking targets of opportunity much easier. As with the early days of flight, air power roles expanded; reconnaissance, followed by ground attack, followed by aerial combat. Each progressively more cognitively challenging.

The Predator is a product of the wars of the day, flying counterterrorism and counterinsurgency missions against determined, but poorly equipped adversaries. Western forces enjoyed total air supremacy here, with almost all drones lost to malfunctions not enemy fire. Neither the Predator, nor its more powerful successor, the Reaper, was designed to operate in a contested aerial environment, with threats from other aircraft or ground fire. A new generation of drones were designed for less permissive airspace, utilising stealth technologies—like the RQ-170 Sentinel, believed to have conducted surveillance of Osama bin Laden's hideout in Pakistan. This small drone looked very similar to the current generation of piloted stealth aircraft, particularly the far larger 'batwing' design B2 Spirit heavy bomber. But just like the Predator, it was not intended for aerial combat. And like all modern drones, its autonomy was constrained. The Predator is a remotely piloted air vehicle (RPAV), and while little is publicly known about the control of the RQ-170 or its larger successors, it's likely that they too are controlled by humans via data links. Since they're designed for operation in contested airspace where electronic defences might degrade communications, it's also likely that they can carry out pre-programmed missions, and perhaps even adapt to changing circumstances, like deploying countermeasures against missile attack. These though are limited AI capabilities, not far distant from those on a cruise missile or the autopilot computer of a civil airliner.

But more sophisticated computer pilots are coming. Five years hence, it's likely that some of the aircraft supporting our Special Forces mission will be entirely autonomous, and able to defend themselves. In 2014, just as the Chinese navy was demonstrating its capacity to launch and recover piloted jets from the deck of its new aircraft carrier, the US Navy was making its own step-change: the first carrier launch of an uninhabited combat air vehicle (UCAV), the X-47b, another aircraft in the batwing tradition. The X-47 was designed as a testbed for ideas about aerial combat. But that's not how it ended up—in the middle of the programme, the Navy leadership changed the mission—their requirement was no longer for a combat aircraft, capable of operating in hostile environments, but for a refuelling tanker, designed to extend the range of piloted carrier aircraft, like the new F-35C Lightning. It was a telling decision. 2014 was only a few years into the modern deep learning revolution in AI. It's still a year before DeepMind's triumph in Atari gameplay made the front cover of *Nature*. It's only six years since Andrew Ng's helicopter was turning somersaults over Stanford University. Perhaps state-of-the-art AI could not deliver the necessary performance for combat, even in the relatively uncluttered environment extending around a carrier task force.

But then, in 2016, came a startling new development. An AI bested an experienced human fighter pilot in simulated aerial combat.[9] Called Alpha, the AI Top Gun was developed in cooperation with the US Air Force Research Laboratory. Intriguingly, it was built not along the lines of an artificial neural network—like those responsible for so many recent breakthroughs—but on a 'fuzzy logic' design, more akin to good old-fashioned AI. There was, however, a modern twist. To avoid the problematic 'computational explosion' of so many parameters, the AI was refined using a genetic selection algorithm—another loose analogy with biology, which allowed the most 'adaptive' parameters in the model to survive for subsequent generations. Over a series of simulated dogfights, man and machine went head to head. The computer had fewer missiles of shorter range than the human pilot. But that made no difference—the developers reported that the human pilot was shot down every single time. In 2020 another DARPA competition repeated the exercise, but with more competitors.[10] This time eight entrants pitched their dogfighting AIs against one another, before the winner squared off against an experienced human fighter pilot. Once again the combat involved simulated aircraft, in this case F-16s, an air superiority fighter plane designed around their human pilot. The winning algorithm, from a small tech firm called Heron Systems saw off mighty defence contractor Lockheed—an indicator that autonomous combat will be no respecter of established reputations. And in the final, it thrashed the human 5–0.

So, why not just bootstrap Heron's AI into the Navy's X-47b or another pilotless plane? For one, it's a long way from a simulation to the real world, which is altogether less bounded, even in mid ocean. Fighter pilots were quick to point out that simulated conflict and the real thing are rather different. Perhaps, but that's a challenge that DARPA will take up in the next stage of its trials. Such is the pace of development that you might expect to be reading about that contest at much the same time as this book. But that sceptical response from some pilots points to another impediment to autonomous fighters: more than performance is at work when militaries choose their equipment. Crewed aviation has become a central part of American military culture, from the real-life drama of Midway in 1942 to the Hollywood drama of *Top Gun 2*. Fighter pilots form a powerful sub-culture within the Navy, shaping procurement decisions and concepts for decades to come.

The short-term future of piloted flight looks assured. The Navy will get its range-extending tanker—though the Boeing MQ-25 Stingray it's commissioned for the role still looks more like a stealth attack jet than a traditional refuelling aircraft. Autonomous combat drones will soon arrive too, but they'll only deploy as 'wingmen' for crewed aircraft. Each pilot could oversee a mini-swarm of mid-sized drones, operating in loose formation nearby. Already the US has demonstrated the concept, using legacy F-16 fighters in the wingman role, controlled from a cutting-edge piloted F-35 fighter. And now there is a purpose-built demonstrator aircraft, the Kratos XQ-58 Valkyrie. This again is a stealthy design, halfway between the F-35 and the batwing X-45. Other nations, notably Australia and the UK are moving swiftly to field their own loyal wingmen.

Note the conservatism of this vision. The wingman concept calls for human-machine teaming, with the human in close proximity to the machine and in overall control. The advanced F-35 already uses AI systems extensively to shape the decision-making environment for its pilot, and the addition of Valkyries to the mix can be conceived as simply an extension of that blend. The same conservatism is on show in the design for the Air Force's next generation bomber, the B-21 Raider. Again, there's the batwing stealth design, and again, there's a cockpit—crewed flight has years left to run, at least in the minds of America's weapon designers and service chiefs. There's been a key difference in the timelines involved in developing new weapons systems—measured in years—and the pace of change in AI capabilities, measured in months, or even weeks. And while it may be that culture shapes the retention of the pilot, there's also a justifiable need to conservatism: staking everything on technologies whose performance is uncertain would be unwise.

Meanwhile, at sea, similar forces are again pushing in the direction of crewless autonomous systems. Already experimental work has been underway with fully autonomous surface and submarine vessels. And here again there is an advantage in the emptiness of the operating environment. The oceans are certainly more complex than even a modern computer game—with currents, marine life and plenty of human activity besides. But they're still closer to that bounded 'toy universe' than is urban combat in a densely crowded modern city. At sea, as in the air, there's scope for radical experimentation with platforms and force design. But so far, modern designs have taken their inspiration from existing crewed platforms—as with the US Navy's *Sea Hunter* autonomous minesweeping and anti-submarine platform launched in 2016. Beneath the surface, the US Navy has also commissioned five 15m long autonomous *Orca* submarines from Boeing, for use in a variety of roles, including anti-surface and anti-submarine warfare. Though large by the standards of existing crewless vessels, the *Orca* is far smaller than an inhabited hunter-killer submarine, like the Royal Navy's newest *Astute* class boats, at almost 100m. That will limit its weapons payload, but offsetting that, there'll be no need for life support systems, which take space and limit the endurance of the boat. A nuclear-powered submarine could remain at sea indefinitely, but the need to feed the crew in practice limits deployments to around 100 days.

Compared to these developments, AI warbots in the land environment seem less promising. Comedy pratfalls at the Robotics Challenge suggest the scale of the problem. Nevertheless, ground combat robots *are* in development—for example, tracked gun carriers that can operate alongside dismounted infantry. Qinetiq, the British manufacturer, demonstrated one in recent trials with the British army in 2018—albeit operating as a remote-controlled platform, rather than an autonomous warbot. Autonomy also features in automated or active defence systems for armour—detecting incoming rounds and propelling a charge into the air nearby in an effort to intercept it, rather like the Phalanx system on ships, or chaff ejected by aircraft against heat-seeking missiles. And AI plays a part in reconnaissance systems that can identify targets for land-based artillery—for example in computing counter-battery fire against enemy artillery. Then there are tactical air systems that can be used by ground forces. That includes small drones used for short-range reconnaissance, or even attack (when mounting small arms or carrying explosives).

So, while combat will remain human-centred in the near future, a range of weapon systems are in development that will introduce AI elements to combat. Whether that involves AI in identifying targets, as part of networked surveillance systems, or AI platforms that can manoeuvre and fire independently, the warbot is already here. More systems are becoming operational all the time, and the pace of change is accelerating, as the rapid advances made in machine learning over the last decade feed through the slower processes of weapon design, validation and deployment. In some respects, the situation is akin to that of airpower before the Great War—the basic concept already existed, but a period of considerable experimentation was required before militaries figured out the best ways to use the new technology. What's now underway is less of a Grand Challenge, and more Wacky Races for warbots. If the physical dimension of warbot fighting power is becoming clearer, the conceptual element is up for grabs. But some features are already emerging.

Faster decisions

Figuring out the best way to use AI is not straightforward, in part because the capabilities of the technology are changing so rapidly. A good starting point for conceptual debate is an appreciation that AI is a technology qualitatively different from its antecedents. It's not a stronger metal, allowing sharper, less brittle swords or stronger hulls; nor is it a cantilever system, allowing the delivery of projectiles at greater range. It's not gunpowder, rocket fuel or a nuclear reaction—each of which greatly increased the destructive power of those projectiles. Nor is it a technology like the lateen sail, sextant, chronometer or steam engine—each of which increased fighting power by enhancing the manoeuvrability and range of warships.

Instead, AI is primarily a decision-making technology. Its effect is on the *nature* of warfare, insofar as it alters the long-standing human psychology of the decisions made in combat. The simulated dogfight between Alpha and the veteran combat pilot illustrates the point—there was something about the inhuman quality of Alpha's decision-making that led to victory. Materially there was nothing special about its (simulated) aircraft; in fact, its weapon systems were worse than the humans. What was it that gave it a telling advantage? As Maverick once told Goose, it was all about the need for speed.

One of the most imaginative strategic thinkers of the last century was the American Air Force colonel John Boyd. His lasting legacy for generations of trainee pilots and War College students is the OODA loop.[11] The letters stand for Observe Orient Decide and Act. Boyd had watched first-hand as US pilots in the Korean War trounced their Soviet rivals, despite flying very similar aircraft. The difference, Boyd conjectured, was the thick struts in the canopy of the Soviet aircraft, which impeded the pilot's vision—at least compared to the American aircraft's bubble canopy. That marginal variation resulted in a dramatic advantage—perhaps as many as five Communist aircraft downed for every American loss. Boyd watched that kill ratio decline markedly during the Vietnam war, albeit that most American aircraft were lost to ground-based missiles and artillery. It gave him an idea: the secret of success in combat was simply to cycle your decision-making faster than an adversary. While they were still taking in the view around them, and figuring out their own position on the battlefield, you had decided, acted, and were already observing the new situation, ready to cycle again. It was, essentially, dogfighting cybernetics—the constant interaction of an agent with its environment; in this case a potentially lethal, adversarial environment.

After Vietnam, there was an increased emphasis in the USAF on aerial combat. Not only did the services set up dedicated dogfighting programmes, like the famous Navy 'Top Gun' school at Miramar air base, they changed the planes they were using too. The air war over Vietnam was fought with so-called 'tactical' fighter bombers, like the USAF F-105 Thunderchief. But this aircraft had been designed with a different primary mission in mind—flying low and fast into the Soviet Union to deliver nuclear weapons in the event of the Third World War breaking out. Jinking around the sky in aerial combat with North Vietnamese MiG 25s was another matter. Why use the Thunderchief? The decision reflected years of cultural dominance within the USAF by Strategic Air Command, charged with the primary mission of deterring the Soviets with nuclear weapons, or at a pinch, executing the plan to destroy them with a massive nuclear assault. Dogfighting was not a large part of this scenario, and that shaped both the ethos of the Air Force and the weapons systems it favoured. In similar fashion, the veneration of combat pilots is shaping the preferences of air forces today, hence the cockpit on the B-21, and even on the RAF's Tempest concept for next generation fighter aircraft—which won't enter service for at least two decades.

Alpha and Heron backed up Boyd's suggestion—decision-making speed matters in dogfighting, and AI is stunningly fast. Speed might not always be the most important factor in tactical combat. But even if not, there may be scope for AI to make an outsized contribution via its impact on decision-making. Perhaps instead it's the ability to spot targets hiding under cover—to recognise a face or a particular gait in surveillance footage captured by a UAV in ultra-high resolution, real-time footage, and processed by the algorithms of the Pentagon's Project Maven. Perhaps it's the ability to sift through mountains of data, looking for a small but telling nugget—like the supercomputers of the NSA trawling relentlessly through electronic intercepts of mobile phone communication looking for keywords; or the hydrophone detectors lining choke-points on submarine routes, waiting for the subtlest sounds of a cavitating screw. Many tactical activities are susceptible, at least in part, to the comparative advantages of deep learning AI. Where there are masses of labelled data on which to train an algorithm with the techniques of supervised learning, as with image recognition, AI has potential. So too where there is scope to rehearse an activity endlessly, like a simulated dogfight, applying the techniques of reinforcement learning, or genetic algorithms.

Warfare at cyber speed

AI's speed will certainly be important in one vital aspect of tactics—the struggle to secure the information on which modern warfare depends. Cyber warfare combines all the elements at which AI excels—blistering speed, pattern recognition, scope for machine learning. Information warfare is wholly reliant on computer systems, and that makes it vulnerable to cyber-attack.

Some theorists have suggested that 'war' is the wrong lens through which to view cyber security, since many hostile actors aren't interested in, or capable of generating violent effects from their activities. That's true for example of those looking to extort, like hackers locking computers until a ransom is paid, or those engaged in cyber espionage to steal valuable information. Some cyber attackers are motivated by malice, like makers of revenge porn, using deep-fakes; others by mischief, or the pure thrill of the game—the hacking equivalent of Mallory on climbing Everest—because it is there.

But plenty of hacking *is* motivated by national security.[12] And while it's a bit of a stretch to say that you can kill someone directly by hacking their code, that's also an overly restrictive definition of what constitutes a weapon. British infantry soldiers sensibly refer to their rifle as a 'weapon system'. The rifle is made of interconnected parts that work as a whole to deliver the effect. But the system doesn't stop with the rifle and bayonet. The soldier themselves is part of the system, as is the organisation that put them in the firing position, drawing a bead on the enemy. Boots, bullets, rations—all are part of the system. And we can draw the boundary of the weapon wider still—the intelligence apparatus that tells the soldier where his adversary is hiding. The company that designed the drone flying as part of that surveillance system. Perhaps even the society that produced the soldier and the weapon designer—its schools, its hospitals. This is a pretty expansive definition of a weapon system—like something out of a cybernetician's daydream. But even if we draw the circle more narrowly, the electronic systems on which warbots depend fall squarely within the scope of a 'weapon system'. So cyberwar 'might not take place' as Thomas Rid has argued, at least in the sense of malign code actually being used to carry out violent attacks.[13] But deaths may ensue anyway as the result of hacking—cutting power grids, hijacking healthcare computers and so on. And, more fundamentally, attacks that compromise and degrade weapon systems are themselves having just as much impact on battle as the bullet fired from the Special Force's patrol's personal weapons.

Cyber security is becoming increasingly intertwined with AI, which can be used to automate both attack and defence. One aspect of that is the longstanding dynamic between encryption and code breaking, stretching back to Bletchley. Another is between the computer virus, which seeks to exploit vulnerabilities, and its patch. Both activities are susceptible to AI's pattern matching attributes. Recognising the risks, DARPA launched yet another 'Grand Challenge' competition, this time for AI that could automatically find vulnerabilities in code, so that they could be patched. In 2016 the winner was an algorithm from Carnegie Mellon University called Mayhem, which acted as a 'white hat' hacker—seeking vulnerabilities many times faster than human hackers could manage.[14] Rather like the AI combat pilot, the system employs 'fuzzy logic' as a heuristic to narrow the tree search of all the options that a program could make to manageable proportions. These were then examined for weaknesses. Then Mayhem generated malign code that could exploit the flaw—anticipating what an enemy hacker would do—before creating a patch. The final element was an expert system, using domain specific judgment to decide when to roll out the patch—too early, and the program's performance might slow, too late, and the hacker would break in.

That's as far as the Grand Challenge went. After all, the competitors are the good guys. But the next step is obvious: to hack the hacker—turning defence into attack by using the same offensive element in Mayhem to identify vulnerable code on enemy systems. In the language of strategy, this is anticipatory self-defence, or pre-emption. But one person's pre-emption is another's escalation. When Mayhem triumphed in 2016, DARPA's project director compared the achievement of automated cyber defence to the initial flights of the Wright brothers. There is no denying that it is still early days for AI-powered cyber warfare.[15]

This cyber competition creates new problems for thinking about war. Questions of attribution feature prominently in the debate about cyber warfare. How do you know who has attacked you? But attribution seems not to be the conundrum that many suppose—if high quality investigative reporters can glean insights as to who was responsible for an attack (the US and Israel for Stuxnet; North Korea for the WannaCry ransomware), then the extensive resources of state security apparatus will often be able to do likewise. The Mueller inquiry into foreign interference in the 2016 US Presidential election provided an insight into these formidable capabilities, as US investigators indicted a number of Russian nationals on charges of hacking American computers. The indictment revealed forensic detail about the organisation of Russian hacking activities—names, units, locations of headquarters—though not, of course, the techniques used to acquire that intelligence.

Rather than uncertainty over attribution, the novelty of cyberwar techniques creates uncertainty about what Clausewitz termed the *grammar* of war. He meant, by that, the way in which particular societies understand violence—what is meant by a particular action, and what follows from it. It's one thing to figure out who has attacked you, as with the accused Russians, but quite another to figure out what to do about it. A similar degree of uncertainty attends discussion of how to treat warbots—for example, what should you do if an adversary scoops up your underwater drone in international waters, and carts it off for inspection/reverse engineering? That's already happened, when in 2016, the Chinese navy retrieved an American experimental submersible in the South China Sea, right under the nose of the ship that had launched it. Provocative, yes, but an act of war? Not in the view of the US military on that occasion. That drone was later returned to the Americans, after diplomatic protests.

What about the sort of cyber hacking the Russians were accused of? There's no clear answer—states are feeling their way forward in new circumstances. So Chancellor Angela Merkel could express her outrage on discovering that Russian military intelligence had obtained emails from her constituency office in 2015 via a hacking operation—but it would be remarkable indeed, even negligent, if German intelligence were not engaged in similar efforts.[16] Again, there is an element to this that is far from novel—states have always spied. It's the new *means* of spying that cloud matters—what is the appropriate response? In particular, when does an enemy act justify escalation, and especially escalation that might involve violence more directly?

There are no clear answers yet, except that a nascent 'norm' might tentatively be detected against military escalation in response to cyber provocation. Whether that norm becomes a taboo, like the deep and stable expectation that states should not use nuclear or chemical weapons, remains unclear. But for now, a like-for-like cyber counter-attack seems more likely as a response to cyber-attack or cyber espionage: hacking back, in the way of Mayhem and its covert successors.

Warbots are hard to defend against

If speed is important in warbot warfare, defence becomes challenging. In strategic theory, much is made of the balance between attack and defence, with analysts expending considerable energy to debate how new weapon systems alter the prevailing balance.

Mayhem worked to bolster the defence by using AI to anticipate weaknesses, and in patching them to build resilience. In the language of strategic studies, this is defence through denial. Mayhem and its ilk are the electronic version of a tactic reaching back into human prehistory; build a wall. And for the moment, the cyber wall seems fairly robust. Mayhem patched vulnerabilities before attackers could find them. And state-of-the-art encryption seems to be one step ahead of hackers—that's why some governments want tech firms to include 'back doors' for law enforcement to access encrypted apps, like WhatsApp. But defenders are on much thinner ice in the physical world of warbots.

Clausewitz, in one of his more elliptical contributions to strategic theory, argued that defence is stronger than the attack. It was a striking claim, that he repeated often. But how can we reach such a sweeping generalisation? Surely it depends on the particular circumstances of the moment, including the weapons technologies that each side possesses. To see what he was driving at, and why AI might challenge that, a starting point is to ask what advantages the attacker has in combat. The 'principles of war' provide some hints—the attacker has the initiative, deciding when to move and where. This sometimes gives them the advantage of surprise, as when approaching from an unexpected direction, or using cover to conceal movement. Attackers choose when to move and where. Both are often critical to achieving success. Much of tactics is about concentrating force in time and space, in order to overwhelm the enemy at that point.

Once you see that basic tactical pattern, you see it everywhere—ants swarming larger prey, the silent long-range patrols of chimpanzees into enemy territory, the German panzer breakthrough at Sedan in 1940. Or the sudden battering of a DDOS attack on a computer network, with thousands of requests for responses arriving at once, overwhelming and stunning the enemy into submission. Of course, executing the tactic isn't straightforward—it requires coordination of the various units, and sometimes communication between them as the action unfolds. And it requires good situational awareness, to know where the enemy is weakest.

Conversely, so much of tactical defence is about frustrating this offensive tactic. Defenders seek to deny all the elements of tactical offence. They scout and place pickets, to thwart surprise, gaining their own situational awareness. And they deny the attacker the advantage of numbers by imposing some structure on the combat. With structured combat, the advantages of numbers are curtailed. There are at least as many ways to do this as there are ways open to the attacker to achieve surprise. So defenders use terrain intelligently—clearing away cover, occupying high land that will slow an attacker's uphill charge, digging moats and building walls. Defenders use disciplined formations to force attackers to fight one person at a time, rather than in a free-for-all. And they use hardening to ward off the blows of attackers—shields, castles, concrete bunkers. Lastly they use the alternative form of defence—the threat of punishment. They hold forces in reserve, ready for the counter-attack, at the moment the initial assault begins to slow, or 'culminate', in the language Clausewitz used. Attacks culminate as equipment wears out, as lines of resupply lengthen, the further forward the combatants go from their depots. And they culminate as attackers grow weary—vulnerable to counter-attacking fresh troops.

This basic picture of tactics leaves much out. For example, the balance between offence and defence depends also on the sorts of weapon systems each side deploys. Nonetheless, it's a good jumping off point for thinking about the tactical impact of warbots on defence and offence. Why would they favour the offence? Speed is certainly important, but there's more at work than that.

First a caveat—the distinction between offence and defence can be rather subjective, and sometimes depends on the intent with which the weapon is used, rather than its physical capabilities. In the famous Cuban Missile Crisis of 1962, for example, President Kennedy and Premier Khrushchev fell to debating whether the missiles that the USSR had secretly sent to their Cuban ally Fidel Castro were offensive weapons.[17] Yes, said Kennedy, arguing that they posed a direct threat to the security of the United States, and were a dangerous escalation in the already febrile Cold War. No, said Khrushchev, they were just there to protect Cuba from the altogether realistic threat of American invasion—after all, hadn't American trained rebels tried to do that only a year earlier? Wasn't the CIA even then engaged in plotting to assassinate Castro? The new missiles would make America think carefully about any invasion. And, to be fair, that's exactly what they did—Kennedy vetoed an invasion to destroy the missile sites for fear that some would get off a salvo before they could be destroyed. So both men could make a plausible case. The lesson: in the end, whether a weapon is defensive or offensive is a matter of judgment that is sometimes rather fine.

With that caveat in hand, it's time to explore the implications for our warbots. Why should offence dominate? This debate is already playing out in the scholarly literature on autonomous weapons, with advocates arguing both ways, and some even having their cake and eating it by arguing that the balance shifts as the number of weapons increases.[18] As ever, academics fail to agree. The challenge is particularly acute because of a distinctly novel feature of today's warbots. While autonomous weapons and AI systems have been around for more than half a century, the recent surge in deep learning AI creates the possibility of swarming behaviours, and this more than any other feature, will likely have an impact on warbot tactics.

The swarm attacks!

Swarms don't require a particularly sophisticated intelligence. At least not sophisticated in the sense we often use the term—carefully weighing goals and ways before deciding. In nature, swarming is often an instinctive behaviour. In robotics it's the sort of skilful control challenge that deep learning AI excels at. Above the deserts of the US Navy's vast China Lake research facility, Perdix and the Defense Department recently demonstrated the state of the art. You can watch the exercise on YouTube.[19] Scores of tiny quadcopter drones deploy from pods mounted beneath an F/A-18 fighter plane. They gather together in a loose formation and then navigate to waypoints on a map, as directed by human controllers. The humans were setting the destination for the collective, not controlling each individual drone in the formation—there were far too many drones for that.

As a proof of concept there was nothing new here. Robot swarms had been around for several years, including those in Vijay Kumar's 2012 TED presentation that flew through narrow obstacles, changing formation as they did so, before navigating inside a building, mapping it as they went.[20] That all happened in the same year that Google cracked image recognition with its deep learning cats. All Perdix really demonstrated was the ability to deploy drones from a piloted aircraft—a small technical accomplishment, but also one that strikingly demonstrated the military potential of micro-aircraft working together. Around the same time as the Navy's demonstration, robot swarms featured in a cautionary video about AI ethics. *Slaughterbots* is a fictionalised story about terrorist attacks using a swarm of micro drones, each armed with a small explosive charge that kill by flying at the heads of their targets.[21] Equipped with facial recognition technology, the drones are unerringly accurate and devastatingly effective. The film-makers from the Future of Life Institute were also behind the open letter urging a ban on autonomous weapons that we saw earlier. Swarms featured again in the inventive science fiction dystopia of the Netflix series *Black Mirror*. In one episode the parallel with insects is explicit: scientists have created robot bees, which are then hacked and used as a weapon. Today, then, weaponised swarms are part of the popular imagination and also a nascent military capability.

The tactical implications are profound. Swarms provide attackers with some powerful advantages. The basic idea is to concentrate force rapidly in the precise place where it enjoys a tactical advantage. Before then, the swarm is loosely distributed, which makes it hard to anticipate—what is its likely line of attack? Afterwards it disperses just as rapidly, making it hard to counter-attack—where, in the jargon, is its 'centre of gravity'? The attacking swarm will likely have the initiative, and perhaps enjoy the advantage of surprise—whether that's in timing or approach. And the autonomous swarm should have a decent picture of where the defence is weakest–drawing on intelligence systems that themselves employ Artificial Intelligence.

It's worth reflecting a little more about that intelligence picture here. AI is already used to give commanders 'situational awareness'—as per the US government's Project Maven. More will soon follow, as AI systems begin to task assets as well as analyse their information. The challenge is to make sense of a vast and growing mountain of information ingested by an increasingly dense array of sensors on the battlefield—far more information than human analysts can hope to comb through on their own.

What sort of material? Perhaps the enemy has a smartphone left on, despite orders not to, geolocating their unit when triangulated with social media data. That's already happened during the conflict in Ukraine, between Ukrainian army and Russian backed proxy forces.[22] Or perhaps a Special Forces unit has been recording their exercise on Strava, a popular sports app. That's already happened too: you could, for a time at least, see the circuits painted in the remote desert around secret SF bases in East Africa and elsewhere.[23] AI's ability to find needles in haystacks is formidable. Pattern recognition in large datasets, spotting unusual correlations that humans might miss, and all with tremendous, inexhaustible speed. All then served up to the attacking swarm.

Traditional means of thwarting that enhanced situational awareness will surely still play a part in resisting attack. Defenders seek to deny by camouflage and dispersal. They use the advantages of terrain and the depth of the battlefield to conceal their forces. But there are new challenges to all these methods—too much dispersal makes it difficult to rapidly reinforce the isolated units under swarm attack. And the AI tools of the attacking forces work against those traditional methods of passive defence. DARPA and the CIA were originally motivated to fund AI precisely because of the promise it held for detecting enemy signatures—the tank hiding in woodland, the coded radio message passed between commanders. It's ever harder to hide from modern AI, with its increasingly sophisticated sensors and real-time analysis. Cluttered environments, like densely populated urban sprawl will continue to offer some advantage to the defenders seeking cover-from-view. But a look at state-of-the-art surveillance in China demonstrates the changing intelligence landscape. A range of AI-enabled technologies, including biometrics and facial recognition, allows watchers to cut through the clutter. That certainly has implications both for citizens in authoritarian regimes and those in democracies interested in preserving individual rights.[24] And it also has implications for those looking for targets on the battlefield.

Defenders deceive

The swarm combines decision-making speed and rapid concentration. If there's any hope for the defender, it probably lies in deception. Think of an electronic version of wooden decoy tanks. Some modern AI systems are particularly vulnerable to this, and there's no ready fix. ConvNets are the deep learning networks we encountered earlier that specialise in the recognition of imagery, like those that learned to recognise cats online. Given Google's involvement in Maven, it is highly likely that ConvNets also feature in the Pentagon's imagery analysis systems. Even the most sophisticated ConvNets don't succeed all the time, but by now they are getting pretty good, benefiting from deep neural nets and ever-greater volumes of training data. A military ConvNet scrolling through footage captured from a wide-area surveillance camera should have a decent chance of recognising a decoy tank from a real tank. But not, perhaps, if the defenders adopt a new form of camouflage: spoofing.

In 2019, Belgian researchers exhibited a t-shirt, with a sort of tie-dye splodge of colour on the front.[25] Their goal was to fool algorithms designed to recognise a person—and it worked, thanks to a remarkable vulnerability in ConvNets. It turns out that superimposing electronic information on the top of an image causes the algorithm to make some dramatic mistakes.

And not only that, but to make them with near-certain confidence. The screen is invisible to humans, but evidently not machines. In one famous example, the image of a turtle is treated digitally, and the AI categorises it as a rifle.[26] The reverse process might be more useful for a defender looking to disguise rifles as turtles. Imagine using spoofing not just to hide targets, but to create them—with the attacking swarm concentrating and hurling itself at a deserted hillside which the system is convinced is heavily defended. This startling technical vulnerability is poorly understood, partly because the internal workings of a deep neural network remain something of a black box.

It's unlikely to be a decisive vulnerability. Targets typically produce non-visual signatures too—acoustic and infrared sensors offer plausible alternatives. Still, along with Mayhem-like cyber defences, spoofing should encourage caution in those, like me, confidently predicting the offence dominance of warbots. And it's likely that there will be further vulnerabilities waiting to be exploited by agile defenders. With the arrival of swarming, the balance between defence and attack has swung towards the latter—but perhaps not forever.

Commanding a clone army

If swarms become a basic building block of tactical AI systems, there will be some sizeable implications for military activity. Command and control of a swarm looks very different from that of a human army. Two key attributes of a swarm are decentralisation and mass. Together, however, these make it impossible for a human commander to direct the behaviour of each individual in the swarm. Instead, there will probably be some higher-level coordination of activity, with individual units reacting adaptively to what they sense going on around them.

Perhaps a human commander will direct their swarm *en masse* to a way-point, or issue a collective 'bombs away' signal to approve action. But there are simply far too many individuals for micro-management. That places a premium on stating the mission and rules in a satisfactory way—a challenge we've already touched on with the 'frame problem', and will return to later, when thinking through the ethical dilemmas of warbots.

Military AI theorists distinguish between a human 'in the loop', 'on the loop' and 'out of the loop'. The 'loop' in question is the decision-loop, and usually the decision that theorists are thinking about is the one to kill. If a system has a human 'in the loop', no machine can autonomously fire without the active approval of a person. Alternatively, a person 'on the loop' lacks the capacity to initiate every action, but oversees what's going on and can intercede to stop actions. That could be a weapon release by a single platform, a re-tasking of the whole swarm or anything in between. This is an acknowledgement that supervising every single unit simply won't be possible for a vast fleet of robots. The whole point of the swarm is to saturate a target faster than a human could hope to respond to. A team of humans staring intently at monitors could conceivably leap in at the last moment if one of the drones made the wrong decision—for example attacking a civilian. But not if communication to the swarm were degraded by enemy defences. So, if such oversight slows down the OODA loop, the whole enterprise may lose its distinctive advantage in making decisions faster than it is possible to react against.

Many western militaries advocate 'mission command'—the idea that a subordinate unit is better placed to respond to events than some far distant commander, and should be given the latitude to do so.[27] The commander states their 'intent'—for the swarm, we can think of this as the top-level order: 'go to this coordinate, destroy the enemy units there'. The details of *how* to achieve the commander's intent are left to those able deputies closer to the action. This approach gives attacking forces the sort of flexibility to adjust to the unexpected and to exploit fleeting moments when the enemy shows weakness. It permits defenders to respond to surprise and chaos without paralysing delays while they wait for orders from commanders who have little idea what's going on nearer the action.

For humans, mission command requires trust and skill. It takes training and doctrine, so that everyone has a common appreciation of the problem, and plenty of practice of working together. It requires, in particular, a cadre of intelligent junior officers and NCOs able to take the initiative and demonstrate effective leadership. The AI swarm doesn't need those martial virtues—with implications for the organisation of armies, including the sorts of people they need to recruit, and the structures in which it places them. But it nonetheless demands a computerised version of mission command. Leaders will be less able to reach down to the tactical level as combat unfolds, no matter how well networked the attacking force has become. This means that the basis on which the swarm decides is hugely significant. Two factors matter in particular: the extent to which it accurately *understands* the context in which it is operating, and the extent to which it correctly interprets the commander's intent. If it gets either of those wrong, there is the danger of applying the wrong amount of violence (too much, or too little) against the wrong target.

So the brain of the warbot is, predictably, its key feature. But what attributes can we expect it to exhibit, five years from now, in the Syrian border lands, the heavily defended airspace of the South China Sea, or under the polar icecap? By now you'll be familiar with the broad parameters—AI's speed, pattern recognition and accurate recall. Set against that are its familiar weaknesses—its brittleness and its narrow, domain-specific skills. And also its lack of 'common sense' and the difficulty of representing knowledge to it in a way that we can be confident fully captures our intentions. This puts a premium on stating intent accurately.

There's another important feature: the swarm can share knowledge rapidly, either in training, or (if communications are unimpeded) combat. Once one unit learns something that improves its performance, the others can rapidly be upgraded with the more optimal version. Tesla already does that with its 'autopilot' driving feature.[28] Expertise can then be acquired far faster than for human soldiers and pilots. One problem with that approach is that the clone army would then share the same vulnerabilities as well as strengths. But still, this sort of rapid learning is a boost to mission command—when a subordinate platform does something that works, the others instantly adapt too—with no need for higher level coordination.

Combined arms swarming

Swarms might exhibit all manner of attributes and capabilities, but as with any weapon there will be trade-offs. This is a hardy perennial of combat. Heavier platforms can deliver more firepower, but are slower. Stealthiness protects, but also comes with a penalty on speed, which generates heat. When it comes to choosing among these and other capabilities, what attributes should designers prioritise?

One feature of many swarms in nature is that the individual members are identical, including cognitively. There will still be individual differences, that we might broadly characterise as matters of personality. One starling will differ from another, for example in its startle response. But they think broadly along the same lines: sensing the world in the same way and responding to it using the same evolved mechanisms. Not so in the warbot swarm. Opposing warbot swarms may possess radically different approaches to control, and radically different cognitive trade-offs. Equally, within our swarm there's no need for functional identity. We might have some units more focused on speed, others on accuracy, depending on the specific roles they perform—for example, speed for combat manoeuvre platforms, accuracy for those specialised in ground attack.

Today's armies are organised around the principle of 'combined arms'—the idea is that fighting power is boosted if specialised units act together. Thus, the Main Battle Tank relies on a screen of dismounted skirmishers to tackle enemy infantry armed with anti-tank weapons. And it may utilise scout helicopters to perform tactical reconnaissance, identifying nearby targets concealed on obverse slopes. Dismounted infantry can call on indirect fire support from artillery many miles distant, or mortars nearer at hand. Combined arms is nothing new. Ancient Greek hoplites—heavy infantry, with their signature round shields, bronze helmets and long spear—were accompanied onto the battlefield of antiquity by cavalry and skirmishers, armed with slings. Roman legions used war engines to throw heavy boulders. The possibilities are endless. Fighting power depends on finding the right combination of specialisms. The same is true of our swarm—it's a new form of combined arms, whose members need not be identical. In fact, we might see a menagerie of specialist robots. Birds of very different feather, flocking together. Anticipating the trend, Britain's government-owned Defence Science and Technology Laboratory has already experimented with small swarms of five different autonomous aircraft.[29]

In recent years, defence inflation has driven the acquisition of platforms that are 'multi-role' generalists—that's especially true of aviation, where the 5th generation fighters which excel at aerial combat are also capable ground attack platforms too, with some capacity for intelligence gathering: jacks of all trades. But there are inevitable trade-offs. The F-35 Lightning is an astonishing (and astonishingly expensive) jet, but not as capable a dogfighter as the thoroughbred F-22 Raptor. And certainly not as capable of close air support as the venerable A-10 Warthog, an ugly, slow moving aeroplane, beloved of infantry units under attack for its reassuring presence and nose mounted minigun. The F-35 delivers ground support from afar. Its precision munitions are accurate, but in limited supply. It's an expensive solution to that one combat problem, but then the A-10 is a hopeless dogfighter. Will swarm members be specialised like the Warthog, or a generalist, like the F-35? I predict they will be specialists.

For one thing, there's scope for great variety because there's no crew to protect. Already, the effect of crewless vehicles on design is apparent—but more radical changes are likely. The autonomous systems we know about today still look rather like modestly modified versions of their crewed predecessors. So, we saw that the X-47b shares a batwing design with the piloted B2 Spirit. It's smaller than the heavy bomber, but still recognisably on the same scale. But this need not always be the case. Weapons platforms have been designed around humans. They need to accommodate their bodies, be ergonomically designed so that humans can control them, and must protect humans from attack. So, the F-35 uses stealth, a Main Battle Tank uses heavy armour and active defences.

Being crewed imposes additional limitations on form. In the air, dogfighting is constrained by the human body's tolerance for gravitational force. Removing the human opens up a wealth of new design choices. Now warbot performance is constrained only by the performance of its component materials, not those of the human body. The changes might be truly radical.

In seas off Virginia in 2014, American fighter pilots witnessed a strange phenomenon—a UFO, moving at blistering speed, before abruptly changing direction on a sixpence. The video footage from their cockpit cameras leaked to the internet, fuelling online discussion about alien spacecraft. Eventually, in 2020, the Pentagon itself re-released the footage, which you can watch online. Years later, one of the pilots described what he'd seen to *The New York Times*. It was mystifying:

> We have helicopters that can hover. We have aircraft that can fly at 30,000 feet and right at the surface. But combine all that in one vehicle of some type with no jet engine, no exhaust plume.[30]

Three explanations seem possible—perhaps there really are aliens among us, responsible for this and other UFO sightings captured by jets. Or perhaps there was a glitch with the electronic equipment in the cockpits, tracking the strange contact. And then, when the pilots came close enough for visual identification, perhaps it was something on the head up displays, or a trick of the light. Both these explanations are exceedingly unlikely. The third explanation is more plausible—that the UFO is an pilotless experimental aircraft, part of a 'black programme' of highly classified research. The same sort of programme gave us the otherworldly SR-71 Blackbird, F-117 Nighthawk and B2 Spirit. In their time, these too looked like alien designs, and boasted otherworldly performance. None were publicly acknowledged until long after they became operational. This third possibility was the view of the two airmen, who noted that the sudden changes of direction were far beyond the tolerance of a human crew. 2014 is only six years on from Ng's helicopter breakthrough. And hypersonic engines remain to this day rather experimental. Whether or not that's the right explanation, the episode makes two points—that lots of what goes on in warbot development will happen outside of public awareness. And second—the attributes that deliver an advantage in combat today may not be those of tomorrow.

Research on stealth fighters attests the point. The distinctive advantage they had was relative invisibility to radar defences. To achieve that, however, other performance attributes were traded off—notably speed, manoeuvrability and payload (since munitions needed to be carried internally in order not to compromise the smooth radar deflecting lines). Stealth still carries value for autonomous systems—as the RQ-170 reconnaissance drone demonstrates—when snooping unobserved is the goal. But it may be relatively less important without a pilot to protect. If AI opens new opportunities for speed and manoeuvre, these then may be more useful in air combat. Hypersonic weapons that are controllable in flight may be able to outpace the surface to air missiles seeking to intercept them. Forward thrust vectoring may allow platforms to execute last second manoeuvres to avoid incoming defences.

Quantity has a quality all of its own

Another reason for expecting variegated swarms is the shifting balance between quality and quantity. The signature feature of a swarm is quantity. But in recent years, defence acquisition has stressed quality—meaning smaller numbers of exquisite platforms, like the F-35. Nowhere is that more apparent than in the procurement of aircraft, but other platforms are affected too—

navies today have far fewer submarines and frigates than previously. In part, the trend is driven by survivability of the crew. Societies have become more risk averse about putting citizens in harm's way. But the main driver has been the competitive pressure of deriving a marginal advantage in combat—weapons systems have become more complex, requiring intricate designs and exquisite materials. Modern military equipment is so expensive because the search for marginal advantage requires costly research and unproven techniques. It takes time and money to mature new, unproven technologies. These rising development costs have in turn driven down production runs, which as a consequence further drives up the unit cost. Hence the modern trend for multi-role aircraft and modular land combat systems.

With the F-35, these factors meant a concurrent design and build—the aircraft evolving, even as it was constructed. And that created scope for the required capabilities to change too, so that Lockheed's designers were chasing a constantly moving target. It's a ruinously expensive way to proceed for all but the wealthiest nations. To share the burden, the US joined with eight allies in the F-35 programme—nonetheless, the price per aircraft remained eye-watering, so that even wealthy partners like the UK could only afford a modest (and progressively diminishing) slate of orders. Now, however, the shift towards autonomy might bring about a radical overhaul of defence procurement, away from this low-volume, long-lead time model. Technology, most especially AI technology, is developing just too swiftly. With a marginal qualitative advantage in AI proving decisive in combat, as when Heron saw off Lockheed in the dogfighting trials, states will need to stay very close to the rapidly advancing bleeding-edge and be prepared to turn to new, less established suppliers. As one American general put it, militaries will have to become comfortable with using disposable equipment.[31] The state of the art today might be obsolete next month.

There are some hints already that a shift along these lines is underway in America. Rumours have been circulating in the specialist aviation press about an advanced, next generation fighter, with capabilities far beyond the not-so-venerable F-35.[32] And the step-change implied by a new 'generation' was delivered not in decades, but in months. That would be a truly remarkable departure from the norm, and I suspect something else is going on. Rather than a single all-new aircraft, the Air Force's Next Generation Air Dominance Programme is more likely a suite of collaborative capabilities than a single fighter jet. DARPA's LongShot programme offers more hints of what's coming.[33] It recently solicited for an air-launched uninhabited vehicle able, in turn, to employ air-to-air missiles. In this imagined future, the physical parameters of a platform are still important, but the revolutionary capabilities lie in the decision-making system, and it's this that must be upgraded continuously.

Soon, then, we can expect very rapid development, validation and fielding of AI capabilities, at least in those countries able to make this challenging transition. This, combined with the idea of swarming, will entail a very different approach to balancing quantity and quality. And perhaps also a very different approach to using the platforms in combat. Already we've seen that the pilotless swarm has many more aircraft than a conventional air wing. AI overcomes the bottleneck of training pilots, which is both expensive and time consuming. Few humans anyway have the requisite skills, so fighter pilots are usually scarce, even before hostilities begin. Survivability in combat is vital, shaping appetite for risk—and influencing both the concepts and platforms that are developed. But with AI, survivability is less central. If you don't need to protect a platform at all costs, then not just stealth, but also armour and other defensive aids become less important. These typically add to cost and compromise the platform's speed, manoeuvrability, range and endurance. So, instead of a few exquisite, piloted platforms, AI means you can create a superabundance of clones, and saturate defences, rather than suppressing them, or avoiding them. With enough attackers, you know that some will probably make it through, and that it doesn't matter much if many individual units are lost whilst soaking up the defence's attention and firepower. You might deliberately factor in some disposability. At the extreme, why bother getting any back at all?

The Perdix swarm was only a demonstration, but micro drones that size won't have the endurance to travel far for recovery, and the mothership that launches them—F/A-18 or otherwise—might not have the capability to retrieve its flock, or be prepared for the risk of hanging about. One approach is to treat the units as 'kamikaze' drones—essentially loitering munitions that can persist in the battlefield before attacking at the appropriate time and place. The Israeli manufactured Harpy drone demonstrates the potential—able to patrol a predefined area looking for its designated target before swooping down and detonating on it.[34] Even closer to our Special Forces patrol is the use of loitering grenades—ultra tactical weapons that look like micro drones you might buy from Amazon, but that carry an explosive charge underneath. Commercial drones have already been used in combat, not least in Syria and Iraq. There was nothing autonomous about these remote-controlled aerial bombs, but the next generation will be, like the Spike Firefly loitering grenade from Rafael or the Hero-30 from UVision.[35]

The Harpy is solitary. And both tactical grenades still need human input before they detonate, at least for now. As yet there's no operational swarm of loitering munitions anywhere on earth. But there's no reason why there won't be. For a glimpse of the possible, return to nature and the murmurations of starlings, or the swooping of gannets hunting sardines. The way they orient around each other, blurring the line between collective and individual, is hypnotic. The gannets offer another interesting glimpse of changes wrought by warbots. Formidable divers, they fold back their wings at the last moment, spearing down into the ocean on the sardines below, in a battle of swarm versus shoal. Could it be that the warbot of the near future will move from one domain to another? More broadly, nature has offered inspiration to roboticists for decades, whether the sinister canter of a headless Boston Dynamics quadruped, or the graceful flapping of a flock of electric blue robot butterflies. This experimental phase, where concepts and platform designs for warbots are proliferating, is a robot menagerie.

The best combination of traits in our warbot army remains undetermined. Even if we allow that the swarm is a potent way of using tactical autonomous weapons, it's not the only possibility. The batwing shape of the B1 features not just in the design of the X-47b, but also Dassault's experimental nEUROn pilotless drone, and in BAe System's Taranis. Russia's *S-70* Okhotnik by Sukhoi is cut from the same cloth, as is China's Hongdu GJ-11 Sharp Sword. The common features of stealth and autonomy combine to produce very similar looking craft. And while the performance of each variant remains unclear (stealth relies on the surface skin as well as the shape to deflect radar signals), the aspiration is clear. These aren't disposable swarming drones that will rely on saturation and manoeuvre. Rather they work to penetrate denied airspace clandestinely, just like their piloted equivalent. The attraction here is not so much a new concept of operations, but a way of mitigating risk—perhaps even at the expense of performance. Autonomous platforms also promise more endurance and range—the rationale for both the *Orca* submersible and the Global Hawk surveillance UAV.

So swarming isn't the only possibility for warbots, but the logic behind it is compelling. Unpiloted swarming creates new possibilities. By combining mass with speed and accuracy in new ways, warbots permit startling manoeuvres at impossible speeds, to arrive precisely at a single weak point. It's those capabilities, resting on the strengths of connectionist AI, for search, pattern recognition and learning that promise a new era in tactics.

There are many possible ideas to work through in the phase of experimentation now underway. In the air, the Valkyrie 'wingman' and DARPA's LongShot programme suggest the possibility of a blended approach with crewed and autonomous platforms swarming together. At sea, meanwhile, crewless platforms could work as mobile hydrophones, drifting together in loose formation over large areas of ocean, listening for submarines. Or *Orca* motherships could deploy autonomous loitering micro drones to work as a mobile minefield for unsuspecting shipping. A maritime Harpy could work as a loitering depth charge, or converge on a submerged target, transforming in an instant from aircraft to acoustic torpedo. On enemy beaches, spherical warbots deployed from motherships could roll into shore and thence up the sands towards enemy bunkers, before detonating. That would help marines tackle one of the bloodiest challenges in modern warfare—the amphibious assault.[36] In the end, swarming is a weapon of brute force, and that is one of the distinctive strengths of modern AI.

From big and slow to fast and light

Consider one facet of the tactical puzzle—the use of maritime aviation to strike targets on land and sea. For decades after its stunning debut at Pearl Harbor, maritime power has focused on the aircraft carrier. Carriers project power far from home. They create an obvious presence, signalling resolve or escalation, as the moment requires. The carrier task group is a formidable adversary, and heavily defended, in layers extending hundreds of miles around the ship itself.

But why build aircraft carriers today? They are large, slow moving and extravagantly expensive, with a shelf life measured in decades; everything the swarm is not. Their air superiority fighters and close-in defences are vulnerable to speed and saturation. And the swarm itself need not originate from a carrier—it can just as easily assemble from smaller, more numerous and distributed motherships—perhaps uncrewed themselves. If the swarm is disposable, these motherships might look a little like the attack submarine of today, armed with scores of cruise missiles, more like gunboats than floating runways. And with no carrier to defend, the role of the traditional surface warship comes into question—the air defence destroyer, the anti-submarine frigate; all are needed today for fleet protection. With no need for humans on board, perhaps the balance of maritime combat power switches instead to submarines—which will be faster, and harder to track, even with the benefits of AI.

It may be that the carrier fleet and the heavy bomber are destined for obsolescence, along with other legacy platforms that pack great individual punch, but that are poorly suited to the new, more agile tactics of the swarm. In this vision, the future of combat is a battle of intelligent missiles. Some may optimise long range, some blistering speed. Some are solitary, while others work together in vast cooperative swarms.

Warbot culture

Armed forces are sometimes slow to recognise change. We already saw that air forces still include cockpits in their next generation designs. Conservatism has a place in military culture—tradition and hierarchy have proven their worth amidst the chaos of battle, working against panic and disorder. But conservatism militates against adaptation, especially in times of rapid technological change, when the implications of new weapon systems are uncertain.

Part of that is down to particular service culture and predilection. Another naval example is illustrative—the development of nuclear propulsion and submarine launched ballistic missiles created a potent new weapon, which soon became the cornerstone of American, French and British strategic deterrence. But not without some internal opposition. Within the navy, there was scepticism from surface warfare specialists.[37] And from strategic air forces, there was a reluctance to cede their leading role in deterrence. Some air power theorists even swung away from the established nuclear doctrine of targeting enemy cities to one focused on enemy forces simply because the new submarine launched missiles were too few and too inaccurate for that sort of mission.[38] The cart, in the shape of the B-52 heavy bomber, was driving the strategic horse.

Military cultures are embodied in people, their equipment and concepts—all may be resistant to change. People remain within and develop loyalty towards their specialism; there's a typically single entry-point, so that professional soldiers, sailors and aviators are recruited at a young age. Then there's the celebration of the combatant—in the US Army much is made of the warrior ethos. The US Marines inculcate recruits with the idea that everyone is a 'rifleman' first, whatever their ostensible trade. In air forces, the status of the drone operator lags that of the fighter pilot, despite them wearing 'wings' insignia on their uniforms, and suffering high levels of combat stress.

Change is possible, of course. There's an efficiency logic to warfare—if you don't keep up with the best kit and ideas, you risk national security. That's what drove previous technical changes, like the adoption of steam engines and steel hulls in the Royal Navy, despite centuries of tradition and vested interest in wood and sail. It drove the creation of independent air forces, when conceptual thinking called for 'strategic bombing' of key targets far behind enemy lines. Those air forces stimulated new intellectual thinking about war, as happened in the RAND Corporation of the 1950s and 60s, where much thought went into theorising nuclear war and designing forces accordingly. And now, there is nascent change in western armed forces in response to advances in AI. The US Defense Department has worked to deepen its existing ties with Silicon Valley, to accelerate the development and adoption of new technologies. Notably its Joint Artificial Intelligence Center (JAIC) has produced a stream of thinking on operational and ethical implications of AI. New suppliers and more rapid procurement cycles promise to shape up sclerotic defence acquisition processes. Yet change will be challenging, even for the United States with its long-established blend of defence and technology research.[39] The challenges will be profound elsewhere too. China and Russia are both eager to embrace AI technologies for national security purposes. Both invest large sums in defence research and have begun to field sophisticated AI systems of their own in recent years.

But it's not straightforward to translate aspiration and investment into capabilities. Both China and Russia have very different strategic cultures from America's, and it remains far from certain whether their ambition and drive for military AI will reap the same dividend. Neither state has America's winning blend of entrepreneurship, strategic government investment or university research culture. And both suffer high levels of corruption, including in research.[40] In China, as elsewhere, cultural and organisational factors shape and may constrain the development and adoption of battlefield AI. China's army has long relied on centralised, hierarchical control, reflecting its Communist heritage. But that pushes against the sort of devolved 'mission command' that is the warbot's signature. Its weapon systems, meanwhile, often emulate types developed in the US—aircraft in particular often look uncannily like American counterparts a generation or so ago. Espionage and savvy business investment have allowed China to emulate technologies developed overseas, and perhaps in some cases producing genuinely world-leading capabilities. But emulation does not deliver cutting-edge capabilities. And acquiring modern technology is a different challenge from implementing it as part of an effective warfighting system. The challenge for China today is to innovate genuinely novel systems. Russia meanwhile invests far smaller amounts in both defence and AI research than the two dominant world powers, notwithstanding Vladimir Putin's striking claim about AI mastery leading to world control. The USSR was slow to the computerised military revolution of the 1970s, and while it's been very active in employing computers for intelligence gathering and disinformation, there's been little evidence of any comparable AI acumen. Russia's great challenge is translating its industrial-era defence base into a post-industrial one that thrives on rapid innovation and creativity.

Still the efficiency logic is driving the acquisition of AI in China, Russia and elsewhere—such is the impact of autonomy on fighting power that no one wants to be left behind. If that means re-organising armed forces to better deploy warbots, so be it. Already China has deployed autonomous boats and active defence measures on its armoured vehicles. It is developing automated battlespace management tools, and its extensive use of algorithms in domestic social monitoring give its national security effort plenty of experience and data. China has tech giants, AI start-ups, and huge government funding. It may lack America's AI pedigree, but its ambition for warfighting AI is undeniable. The rapid modernisation of its economy and its armed forces alike seems to have fostered a willingness to experiment, and to cast aside outmoded technologies and practices.

And more radical changes may soon follow, in China and elsewhere. Consider the way in which armed forces are organised today—split into three services, each specialised in one domain of war, on land, sea and air. This traditional structure is already fragmenting, with organisations tasked with cyber and space warfare developing distinctive identities within the defence establishment. AI may continue the trend. After all, the defining feature of the warbot is its code, not the platform. Automated warfighting systems will reach across the traditional domains. While warbots will still be designed for specific environments, the work on concepts, procurement, maintenance and upgrades, cuts across service lines. Perhaps rather than traditional domains it will make more sense to organise militaries along functional lines—specialists in reconnaissance, attack, mobility, planning and so on.

The basic building blocks of the three services will change in other ways too. The drone 'pilot' is the leading indicator of changes to how humans will fight as automation increasingly dominates combat systems. Autonomy pushes humans out of the loop, and away from the action—and perhaps the warrior ethos will change accordingly. The relatively small 'teeth' of modern militaries will shrink further. And the bond between society and those charged with fighting for it will become ever more tenuous. In the United Kingdom, with a population of some 68 million, the armed forces number fewer than 200,000 people, or 0.3% of the total. And that includes all the support branches—the actual number of combatants is smaller still. That may have implications for the way in which society thinks about warfare. It may also have an impact on that 'warrior ethos' within the service—perhaps leading to rebalancing away from traditionally venerated attributes like physical courage in combat, moral leadership and command presence. Other skills, like technocratic competence and STEM expertise may become even more highly valued than presently. Skills that an AI cannot master may be particularly important—again, those include an understanding of context and meaning, emotional intelligence and empathy, and also the capacity for creativity and imagination. These are uniquely human skills, and point to the connection between tactical and strategic activity.

The organisation of military sub-units is also ripe for disruption by warbots. The existing structure has deep roots. It reflects the historical need to organise huge numbers of men in battle, to cope with the increasing complexity of combined arms warfare. The split between war-fighters in their regiments and professional staff officers at headquarters is centuries old, a product of Napoleonic Warfare and the Prussian reforms it precipitated. The split between officers—captains of bands and colonels of regiments—is older still, with roots in the feudal system between landed gentlemen skilled in combat, and peasants impressed at time of war.

Farther back, we can even detect an echo of the modern military in our evolutionary history. The modern company and the Roman legion are about 100–150 strong. Some evolutionary psychologists and anthropologists argue that this number has its origins in the size of hunter gatherer bands—the group within which humans evolved our sociability—especially the ability to track important relationships.[41] Who was friends with who? Who should you trust? Understanding and channelling psychological motivations in small groups of combatants is vital.[42] Combat can be a shattering experience, and armies have learned how best to cope with it via a blend of realistic training, discipline and motivational leadership. Today, considerable effort goes into identifying leaders with the right stuff—the physical and moral courage to inspire followers.

None of this is relevant if the combatant is a robot. You don't need a fire team of close comrades pulling for each other. You don't need the inspirational NCO with years of experience and a combat ribbon. You don't need the 'grooming standard', setting out hair length and prohibiting facial tattoos. And you don't need the senior officer with martial bearing and inspiring pre-battle speeches. Science fiction again comes to our aid in imagining a possible future. In the fun blockbuster *Edge of Tomorrow* Emily Blunt and Tom Cruise battle an alien invasion, fighting as soldiers in a conventional-looking human army. The aliens, by contrast, are blisteringly fast, autonomous, cloned and networked—a little like my vision of a warbot future. Cruise eventually leads his platoon to an improbable victory, drawing on conventional military leadership attributes—formidable courage, physical prowess, even his trademark humour in a tight spot. But those skills aren't what tips the balance. The film is a sci-fi twist on another movie, the classic *Groundhog Day*: every time the aliens kill Cruise, he goes back to day zero, and starts over. He and Blunt learn how to win by taking thousands of repeated goes at it—just like a deep-learning AI. Sometimes needs must—whatever strategic culture dictates.

Warbot warbot designers

One way to overcome powerful institutional culture and vested interests is if machines themselves play a part in force design. Already some developments hint at what's possible. DeepMind is experimenting with algorithms to design algorithms. Immersed in a particular environment, the new algorithm looks for features it could use as a basis for other reinforcement learning algorithms.[43] It's meta-learning, or learning what learners need to learn. Once the master algorithm has figured out what to look for in each environment, that information is used to train specialised algorithms which quickly achieve superhuman performance in each game. This is a form of unsupervised learning—of an algorithm making sense of its environment from first principles, rather than by working backwards from a set of laboriously labelled data. As with its earlier research, the test bed for DeepMind's research was a suite of classic Atari arcade games—suggesting that we are still a long way from an AI designing a new submarine.

Other deep learning approaches will doubtless play a part here too, like Generative Adversarial Networks, which improve performance by competing against one another—a very loose analogy to natural selection. If they're anything like other neural nets, the applications of these techniques to weapon system design should produce some surprising and counterintuitive results. An early example is an otherworldly looking satellite antenna designed by NASA way back in 2006, using a similarly Darwinian approach to AI. The intricate antenna that emerged was more effective than the human design, but not one any human would have created.[44]

Designing weapons and their components is one thing, but AI will be used to consider the whole operational environment. After all, it's the knitting together of combined arms that creates fighting power. Again, we can see forerunners already here—DARPA's Deep Green research project from the early 2000s is a good example. This was intended as a planning tool for military staff officers to rehearse operations and perhaps even to allow automated decision-making.[45] Not much has been heard of it lately, probably because the hyped capabilities far outstripped the reality of AI then, whatever the moniker 'Deep' suggests. More capable AI systems today will permit more realistic modelling of tactical environments, hence DARPA is currently experimenting with swarm tactics in a virtual environment, a sign of things to come.[46] In the UK, meanwhile, the RAF recently stood up an experimental swarming drone squadron and began working on simulated environments in which to test conceptual ideas.[47]

Now the race is on to develop the best mix of weapons and the best concepts. If there really is a first mover advantage, then perhaps speed is the overarching factor in decision-making. Or perhaps it's accuracy. Does it matter if you are 2 percent more accurate in your decisions than the enemy system, if that system is 2 percent faster than you? Working out the balance between these attributes will be a matter of experimentation in wargames and simulations for years to come. As with all such scenarios, the moment of truth will only come when one system squares off against another. One thing's for sure—an autonomous designer won't double down on established systems and concepts because it has spent decades imbibing the prevailing organisational proclivities on its route to senior command.

i-Battle gets underway

Back to the desert town. Our Special Forces team has some robotic armour support. Much smaller than a Main Battle Tank, the Estonian designed Milrem Type-X Robotic Combat Vehicle looks pretty much like one, even though it has no crew to protect.[48] Tracked, with a turret and a 50mm main gun—it's an indicator that conservative forces continue to influence weapon design, even in times of rapid technological change. But at only 12 tonnes, the Type-X is air transportable, unlike a real tank, which has allowed its rapid delivery here and its reassuring presence in a defile behind the squad.

The team has deployed some other AI systems around their position. Overhead, there's persistent surveillance from a Reaper drone. This one is flying a racetrack loop without human control—the expanded drone fleet has long surpassed the ability of Air Force recruiters to find and train human pilots. The Reaper is hoovering up electronic information from the battlefield, and feeding it into a vast, and largely automated analytical system. Its 'stingray' system is monitoring cell phone calls made in the town—looking for information from enemy spotters, often unarmed, sometimes children, who will relay information on our team's position—and searching for cell phone activated bombs. Its Gorgon Stare camera is filming the whole town in ultra high-definition. Thousands of miles away, the computers of Project Maven are tracking multiple targets in real time: individuals previously identified from their biometric data, facial recognition and even gait analysis.

At US Special Operations Command in Tampa, all this information is being fed into a huge, simulated 'synthetic operating environment'. Created initially as a side project of a video gaming company, American access to Project Improbable has been provided by the British Ministry of Defence after it proved its value in their wargaming exercises.[49] Now it's being trialled in a real operation. Masses of data on the town has been fed into the system ahead of the deployment; some highly classified intercepts, some human intelligence from local informants, and plenty of open source information too—about the town's electricity grid, transport network and food supply, along with information scraped from social media. There's no CCTV here, and little census or health care information, but if there were, that data too would have been plundered. Improbable allowed tactical commanders to rehearse the deployment. What would happen if they put a small team into a district at the northern end of town, looking for an important enemy leader? They rehearsed various scenarios and weighed the odds. In most scenarios, the squad get their man, and so the mission was approved. Now Improbable is generating crude predictions in real time about what's coming next. The system isn't perfect—because AI still doesn't really understand anything. But as an aide to human intelligence analysis of the battlefield, it's considered a useful addition, acting as a 'red team' to test the judgment of the HQ operations cell.

As they prepare to advance towards enemy territory, the team deploys some tactical loitering munitions. Small, lethal UAVs, these Hero-30s are one small firmware update from operating fully autonomously, like the infamous *Slaughterbots* that hunt down people via facial recognition AI. But the US remains committed to human control, even as evidence comes in that a rival superpower has been experimenting will fully autonomous munitions in its struggle against an internal separatist movement. Instead, the SF patrol's grenades require a positive decision from the controlling NCO to kill—there's a man on this loop, even if he'll be trusting an algorithm to pick the target. Far overhead, in the pods hanging from the wings of two F-35s, thousands of lethal micro drones are stacked together, waiting to be launched. If something goes terribly wrong and the patrol is in danger of being overrun, the swarm will kill anyone carrying weapons within a 500m radius of its positions. Encrypted identifiers embedded under the skin of the A-Team signal that they are friendlies and pioneering combat cyborgs. At SOCOM, a watching lawyer sits with a commander, 'on the loop' and poised to call off the lethal swarm.

Moving off, the patrol powers up the 'smart shooter' targeting system on their rifles.[50] By now a mature technology, the 'augmented reality' of the sight makes corrective calculations based on wind and temperature data. It adjusts for body movements by the rifleman, and projects a red dot in the sight of where to aim, based on the target it determines he's trying to hit. The human can press the trigger, but until the barrel passes through the optimal point, the weapon system won't fire. And when it does, the second generation EXACTO bullets loaded in the magazine will alter their course in flight, to track a moving target.[51] As the squad advances cautiously, the Milrem's hybrid-electric engine quietly hums to life and it starts down the road behind them. Its heavy cannon covers their advance and its battle management software fuses possible target information fed from two Anduril Ghost mini-helicopter drones as they move autonomously down the block. Important information is pushed onward to the soldiers, who peer down from time to time at the fold out screens mounted on their chests. Over the horizon at the Task Force's forward operating base, a Cormorant autonomous robot helicopter also whirs to life, standing by to evacuate any casualties.[52] The Americans bought that off the shelf from its Israeli manufacturer as an urgent operational requirement.

The technology here extends far back from the battlefield and the SF squad. It doesn't float on top of society, but is integral to it—shaped by it, and shaping it in turn. Autonomy is at the sharp end, in the bullet that changes course mid-flight. But it's also far in the rear, rehearsing the mission at Improbable. AI even played a part in generating the force, when army algorithms sent recruiters looking for the best SF recruits, and as algorithms tracked their health data in the run up to deployment.

Fears of a dystopian future with pervasive societal surveillance by AI have not been realised in America, five years from now. But AI nonetheless acts as an agent of 'paternalistic nudge'—shaping the lifestyle decisions of millions of Americans, including our A-Team. That makes many in American society uncomfortable. There's no doubt that widespread adoption of AI is changing the relationship between the government and people. And it's altering the fabric of American capitalism too—companies prosper by collecting information about their customers and targeting products accordingly. That's true whether they're selling health insurance or coffee. There are some unforeseen consequences. Our A-Team really wanted access to their target's cell phone data, but he's using an iPhone, and Apple have refused to give it up, even after a FISA wiretapping request from the FBI.[53] The all-powerful surveillance government that activists feared is stymied by the all-powerful technology corporation. In the end, the balance between the two depends on the institutions and the people who shape them. The warbots that America fields will look and behave differently from China's. And those of the

Netherlands will look different too. All that might seem a long way from the front line. But when it comes to i-Battle, the story is about far more than tactics and weapons.

The SF patrol looks futuristic. But as you read this, every part of their kit is already available. It's either operational, or close to it. For those hoping to regulate or even ban autonomous weapons, this is a sobering thought: The AI genie has well and truly left the bottle. Modern military systems are already heavily reliant on AI. We'll return to regulation and ethics later, to design our rules for warbots. But they too are part of that bigger picture—the relationship between society, technology and conflict. Like the weapons themselves, ethics don't just arrive out of nowhere. They're part of a complex dynamic, emerging from society, and shaping it in turn.

So the A-Team advances towards its target. And several blocks away, the enemy, Russian mercenaries and local militiamen prepare to meet them, equipped with some AI of their own. Like the Americans, Russia has been using Syria as a testbed for its AI weapons for many years. Now their unofficial proxies are fielding autonomous machine guns mounted on tracked weapons carriers, along with some autonomous tactical drones. Hundreds of miles away in the eastern Mediterranean, a Russian Kilo-class nuclear submarine armed with 40 Kalibr cruise missiles moves towards the surface, ready to launch if needed. Closer at hand, in the low scrub foothills some 30 miles from the town a detachment of Russian technicians prepares a trailer armed with Chinese made WS-43 loitering munitions, able to hold position for thirty minutes while acquiring their target.

What happens as battle commences? When the shooting starts, it probably won't look much like the swarm-versus-swarm idea outlined earlier. Five years from now, the battlefield will still be heavily populated with human combatants. Ground combat robots will be less clunky than those of today, but there's still a place for infantry in close battle, especially in cluttered urban terrain like this desert town.

As battle unfolds, will the losing side call for reinforcements? Who makes that decision, and what information do they use? Perhaps it's down to the human patrol leader; or perhaps the Millrem's battle management software can task platforms according to some pre-established rules of engagement. Concepts come and go in doctrine circles, but there's one now called 'mosaic warfare' which captures some of this flexibility—a suite of nearby platforms and capabilities are 'offered up' to commanders—ready on standby to be combined in the most effective way at the right moment.[54] But the theory is ambivalent about whether that combining is done by human, machine, or both in concert. This knitting-together elevates the action away from the narrowly tactical matter of battle to the larger canvas of 'operational art'—the connective tissue between combat and the political purpose of war.

We may focus our attention on the capabilities of platforms—the dramatic new technology of an autonomous tank and the loitering munition, the otherworldly grace of the robot swarm. But the real power of the warbot is its place in a system—in the processing of information. The role of the operational commander is like that of an orchestra conductor—weaving together the individual firefights into a coherent whole. In making those combinations, if it does, the focus of AI shifts away from the physical to the cognitive. In combat, AI's signature strengths in pattern recognition, in speed of decision and accuracy of recall are applied to the manoeuvre of objects and the precise delivery of firepower. But that's not enough for the battlespace manager. They need to think in more elevated fashion—what will it take to coerce the enemy? This requires judgment, for example about whether to escalate. It may require creativity about the combinations of capabilities to assemble into the mosaic. And perhaps it requires imagination too—about the possible end points of all this activity. As we pull back from the immediacy of the small battle, the problem becomes less bounded: tactics gives way to strategy.

Is the sort of AI we've discussed so far up to that? You might think, looking at the example of the small SF raid, that there's no need to delegate full operational control to a machine. In the operation we've considered, the human commander sits squarely on the loop, with enough time and perspective to make the relevant decisions. But not every war will look like the ones of the recent past or even the near future. Confrontations of peer rivals, armed with similar fleets of highly integrated, extremely destructive warbots, cannot be ruled out. Their mosaic managers might scale up the destructiveness massively, in the blink of an eye. But can warbots do strategy or are they simply high-powered calculators?

MINDS MAKE STRATEGY

An ace warbot pilot is one thing. It demands relatively simple cognitive skills, of the sort that AI can master. But militaries have a larger prize in mind: AI that can think more deeply about warfare. They are seeking an Artificial Intelligence that can play a role in crafting strategy, deciding how to use force to achieve our goals. It's a formidable challenge. Can it be done?

In the apocryphal tale, we are in the late 1960s, and war is raging in Vietnam. A Pentagon computer has been tasked with analysing reams of incoming intelligence from the field and making sense of it all. 'When will we win the war?' ask the eager analysts. 1964, spits back the computer. Garbage in, garbage out, as they say.[1] So far as anyone knows, there was no such computer, and the story, which was current at the time, simply reflects the mordant wit of those familiar with the desire of the senior leadership to quantify everything. Defense Secretary Robert McNamara was famous for this positivist approach—if you can't measure it, he argued, you can't improve it. This view had brought great success in his years as the youngest President of the Ford motor company, not least in improving car safety by rigorously analysing data on crashes and injuries. Once installed in the Pentagon, he'd earned the ire of the brass by asking tough questions about nuclear strategy and weapons procurement. And now, in Vietnam, McNamara was looking for a way of measuring success in war, that most elusive yardstick.

In the South, where a vicious guerrilla campaign was grinding the Americans steadily towards defeat, McNamara alighted on the now infamous 'body count'—a crude daily measure of how many enemy fighters had been killed. To the north, in the skies over Hanoi, an air campaign was under way, designed to shatter the resolve of the Communist regime to support the insurgents. But how to measure that resolve? A stubborn silence was all that met repeated American entreaties to negotiate. Eventually the Americans opted to measure inputs, not outputs; counting munitions dropped, and sorties mounted. Those metrics were very crude proxies for success and in the end, they weren't good enough. Even measuring them produced distortions—if you were tasked with increasing the rate of sorties, why not parcel your munitions up among more aircraft, rather than sending each out with a full load? If you were counting dead bodies, why not include bodies you weren't really sure were guerrillas? With distortions like that, even a whizzy supercomputer could not reliably anticipate the so-called 'tipping point' where the numbers would turn in your favour. And computers weren't anywhere near sophisticated enough for that sort of job anyway.[2]

But what about now? Could you use an Artificial Intelligence to sort through intelligence and to make sound decisions about the use of force? What might such a machine look like? This sort of AI would be good not just for running the tactical action in combat, like those considered earlier. It would do intelligence analysis, and then it might offer a solution, with some associated probabilities, allowing human strategists to be better informed about the action. It might even overcome some of the biases that humans bring to the analysis of information—as with confirmation bias, when we tend to seek information that backs up our existing beliefs, rather than challenges them; or 'groupthink', when we tend to form a consensus with the small group of decision-makers around us. If a machine were less susceptible to those biases, it might produce better strategic decisions.

AI intelligence operatives

The first requirement for any strategic AI is to understand what's happening. The portents are not good. In the early days of the Cold War, the US rolled out an early warning system, designed to deliver advance notice of incoming Soviet missiles. There was no way to shoot these down, and civil defence measures offered only limited prospects of success. Analysts calculated that millions would die in the initial blasts, and millions more from radioactive fallout in the weeks to come, as a 'nuclear winter' set in. But if the missiles could be reliably detected, at least there was a chance of protecting key decision-makers and of issuing orders to retaliate. A robust early warning system bolstered deterrence, because it strengthened the likelihood that a 'second strike' retaliation would be mounted. But, as Daniel Ellsberg, a Pentagon analyst later notorious for leaking the Vietnam era 'Pentagon Papers' revealed, the American system didn't work. Or, rather, it worked too well—sounding the alarm frequently, even when there was no attack under way.[3] The first time it went off would doubtless have been utterly terrifying. But the tenth? Like the fable of the boy who cried wolf, the system was quickly ignored by increasingly nonchalant technicians. The problem, it seems, was the moon: the ultra-sensitive radar was picking up an echo from its surface and attributing that to an incoming intercontinental ballistic missile. Presumably the boffins eventually managed to ratchet down the sensitivity to a more appropriate level. A similar panic happened on the other side of the Iron Curtain, this more widely known. In 1982 at a time of heightened tension in the Cold War, an unflappable Soviet Colonel,

Stanislav Petrov chose to ignore the alarm from his new state-of-the-art early warning system which signalled that the US had launched intercontinental missiles. Happily, in both cases, humans were resolutely in the loop, and sufficiently sceptical about the ability of technology to interpret reality.

Ellsberg features again in our discussion of strategic AI, thanks to his vivid depiction of his job as a mid-level Pentagon functionary in the early 1960s.[4] In this pre-computer era, the job of the intelligence analyst involved wading through a large stack of papers. In his memoirs, Ellsberg relates how he would arrive at the Pentagon every morning to a tower of paper—raw information from the field, and the work of other junior analysts. His task: to transform the formidable pile into a much shorter one, and pass that on to his boss, John McNaughton, the Department's Assistant Secretary for International Security Affairs. In turn McNaughton would digest the material and produce a shorter stack still, which would make its way to McNamara. The President's Daily Briefing, by the time it arrived in the Oval Office, with its CIA briefer, would indeed be brief. And so a mountain of paper was sifted and ruthlessly culled for the President's attention.

A similar process is at work today, but now the volume of material in the original pile has expanded exponentially. No one can know how much information there is, but we can be sure that it's expanding rapidly. The internet is partly responsible. Think of the many thousands of pages of information that Facebook sends if you ask for everything it holds on you. Now add in all the WhatsApp chats you've ever had, and the Instagram stories. Then there's TikTok, Twitter and YouTube. Every story, every picture, every purchase. What about all the GPS signals your car navigation system clocks up as you drive? With 5G cellular networks arriving soon, the information deluge will increase again, as the 'internet of things' becomes a reality that even determined luddites cannot avoid. Or consider the vast quantity of information in your genome. Now we can sequence that, and analyse it, expanding our understanding of disease and making personalised medicine a possibility. Genetic information offers insights too about personality traits and behaviours. And everywhere we go we leave traces of our DNA—useful for forensic analysis. DNA might even offer a place to store information—a vast biological hard drive. Thus are millions upon millions more bits of information added daily to an impossibly large pile of raw intelligence.

Perhaps AI could help make sense of this torrent of information. Natural language processing and image recognition were, after all, a large part of the attraction for the defence funders of AI research back in the mid-twentieth century. And modern AI is spectacularly good at recognising patterns in huge data sets. It can transcribe spoken words into text, read even the most unruly handwriting, lip-read visual imagery. It can translate language in real time with increasing acuity. Facial recognition algorithms are increasingly adept at recognising individuals, even when the imagery is blurred, or poorly lit. In financial markets, algorithms track stock prices, making huge numbers of trades in rapid succession, in pursuit of the smallest financial advantage over competitors. So intense is the competition that traders seek to position their machines as close to the market servers as possible, to minimise the time latency of signals travelling along fibre optic cables.

Allied to increasingly sophisticated sensors, this sort of AI could be incredibly useful for intelligence analysis and situational awareness. Fusing together information from a variety of sources allows analysts to build up a 'pattern of life', allowing watchers to track and target individuals, and perhaps even to predict future behaviours, much like Tom Cruise in *The Minority Report*. That's the dream of companies that offer 'predictive policing' services to government agencies, even if the reality is rather sketchier. Much of this technology is 'dual use'—employing exactly the same processes that allow Amazon, Netflix and Spotify to offer personalised recommendations. It's all just pattern recognition in large data sets.

One challenge for military intelligence is to get the right data in the first place. Intelligent sensors offer a solution. 'Gorgon Stare' is the code name for an ultra-high definition camera mounted on American drones. The drone sits on station, flying in a racetrack pattern over a city and recording wide angle images of a large area below. Its camera records everything in incredible detail, including the crimes committed beneath. Later, the analysts can spool footage backwards from the crime scene to follow the perpetrator back to their home, or their associates. Or they can spool forward to see where they go next. This technology, now used by city police forces in America, was originally designed for the military, seeking out insurgents in the urban sprawl of Baghdad.

The next step, already underway, involves automating that process. Doing so would help overcome some big bottlenecks in tasking and analysis. With rapidly expanding fleets of drones, the American Air Force, and its ally the UK's Royal Air Force, are struggling to recruit enough operators.[5] The work is an unusual combination of tedium and stress. After tracking a target over days and weeks, perhaps even coming to know their personality, the crew may find themselves abruptly killing them. Drone pilots suffer post-traumatic stress—indeed by some accounts this is as much a problem as for frontline combat soldiers. Perhaps it's the sense of dislocation—killing from an air-conditioned trailer on the far side of the planet, before returning home for dinner. There's also a perceived lack of status, at least relative to their more glamorous airborne counterparts.

Meanwhile, the expanding drone fleet is generating far more video footage than can be analysed by humans. Why not task an algorithm with determining what it shows? This is the goal of Project Maven, the Pentagon's overarching effort to bring AI into the imagery analysis process.[6] Straightforward it is not. AI can recognise some objects, but tracking individuals is rather harder. At the moment, even the most high-definition drone footage cannot employ facial recognition—the angles are all wrong, and the top of a head is much less distinctive than a face. But there's nothing to stop an algorithm from categorising and tracking other features in a drone video, such as buildings or cars. And individuals *can* still be tracked, even if they can't always be definitively identified. Moreover, there may be other distinctive features than their face through which to identify them—gait or body size, for example. Perhaps by fusing the information from the camera with other sources a more reliable identity can be established.

That's the goal in China, where facial recognition systems working with CCTV footage and imagery from police drones are becoming ubiquitous. There are huge ethical dilemmas here—especially in the west, where individual freedom and privacy is prized by many and often enshrined in law. Intelligent surveillance technologies like these challenge the status quo, altering the balance between public and private space. Project Maven caused a stink among employees at Google, which was working with the Pentagon on it. Google eventually pledged not to renew its Maven contract with the military. The protest was a concrete illustration of tensions emerging between concerned researchers in the private sector and the national security apparatus. In 2020, as the Black Lives Matter protests spread through America, more followed. Microsoft, Amazon, IBM and others pledged not to work on facial recognition, or at least to pause it pending rigorous legislation about how it could be used. Amnesty International and other campaign groups called for an outright ban on the technology. But as Google stepped down from

Maven, Palantir, another data management giant, stepped up. It's not clear how far ethical concerns from researchers will impede the development of AI surveillance technologies. If anything, perhaps they'll feed into wider debates about the appropriate social balance between public and private spaces, and thence into formal institutional arrangements.

AI analysts

More broadly than facial recognition, AI as an intelligence analyst is looking for meaningful correlations, or patterns in data. After all, that's what a face is—a series of interconnected patterns. Often patterns are obscured by great volumes of data, and sometimes they are very subtle. Leftfield thinking can reap dividends. One AI found a link between the make of car on driveways and the voting intention of householders.[7]

This pattern recognition is just the sort of thing that AI excels at. Or rather, excels at to a point. Finding patterns using its immense capacity for statistical processing and accurate memory is one thing. But finding *meaningful* patterns is quite another. As any scientist knows, correlation is not cause. Tyler Vigen has compiled a wonderful compilation of meaningless correlations that highlights the problem.[8] US spending on science, space and technology, for example, closely tracks suicide by hanging, strangling and suffocation. A curious coincidence. And the number of computer science doctorates awarded in the US correlates almost exactly with revenue generated by arcades. Algorithms would find these patterns with little difficulty, but the underlying absurdity of them would elude even state-of-the-art AI. It's that familiar problem of meaning. Connections are one thing—grouping things together into categories is something an AI can do, via semantic maps—but intuiting meaning is harder.

There's another problem with AI analysis. Sometimes it's just not that good, compared to human judgment. Consider algorithms that purport to read emotions from human faces. This is an aspect of AI research with potentially lucrative payoffs. After all, an emotionally attuned algorithm should be able to offer companies better insights about what consumers want. It might conceivably be able to assist in therapy for patients. If it were any good, that is. But it's not.[9] There is some evidence for universal, or basic emotions that are shared cross culturally. AI can, more-or-less, label faces that are smiling, versus those that are grimacing. But a more subtle, social emotion, like *schadenfreude*? Not yet.

And prediction on the basis of pattern matching in data is also poor, despite the *Minority Report* ambitions of some designers. One AI system attempts to predict the rates of recidivism among offenders released from prison. Its results are better than chance, but not by much, and no better than the expert humans who currently make decisions about early release.[10] Predictive policing is a reality today—as algorithms crunch numbers on criminal activity in an effort to apply scarce policing resources more effectively. The results though are patchy. And there are valid concerns about civil liberties and racial profiling. Some forces have wound up their programmes altogether.

There are algorithms that aim to assist recruiters by examining the résumés of job seekers and predicting who will be best in the role. With limited success. One problem here is that bias inherent in the training data is faithfully mimicked in the performance of the algorithm—with the result that the HR AI can be prejudiced against women.[11] And then there are the difficulties facial recognition systems have with identifying the gender of black people's faces—again, likely the product of biased training data.[12] If it can't get something that basic right, what chance of building an accurate intelligence picture? If that's not sufficiently ethically troubling, what about another AI whose designer claimed that it could identify homosexual people? Imagine that system being acquired by a repressive, homophobic government. There's only limited evidence that it works: it seems the AI is picking up on extraneous details, like grooming, makeup or spectacles.[13] AI like that is just a modern-day version of phrenology, the discredited pseudoscience whose adherents aimed to identify personality traits on the basis of skull shape.

The bottom line though is that AI can contribute towards building an intelligence picture for strategists. It can process reams of information that would otherwise never be analysed—electronic intercepts, like emails and phone calls; satellite and drone imagery; and the movement of financial markets. But there are many things that even the most adept AI will miss. A rich, human understanding of what information means eludes it. And even the best AI would miss data that never reaches its sensors—the last six inches between the ears of an adversary. That's the very basis of strategy: 'know your enemy', as Sun Tzu, the author of *The Art of War*, counsels. Humans struggle with that sort of intelligence too. But unlike the machine, humans can use their judgment to imagine what the enemy is thinking. After all, aren't they a little like us?

Theory of mind

Once the intelligence processing algorithms have spooled through the available information, what then? The goal for strategists is to use this information to predict what will happen next. Spotting elusive patterns in massive, noisy data sets is one thing, but predicting what will happen on the basis of all that information is another, altogether tougher, challenge. There are stark differences in the ways in which humans and machines go about doing this, which can tell us much about the prospects for AI strategists.

Earlier we saw that KGB chief Yuri Andropov's computer model of geostrategy offered an early warning of the perils of relying on computer predictions. Rather than an accurate assessment, VRYAN's flawed predictions fed Andropov's existing prejudice and paranoia about the intentions of the western allies. Even without the computer's analysis, Andropov might anyway have instructed his spies overseas to gather incriminating evidence suggesting an imminent western assault. But he was wrong, and so too was the computer: a dangerous combination.

And after all, what chance did a computer really have of understanding the real intention of Ronald Reagan, the bullish American President? The job was hard enough for humans. How can we know what the enemy is thinking? Can we even be sure about why we decide the way we do? These two skills—empathy for others and self-awareness—are really complementary. Psychologists sometimes bracket both together under the rubric of 'emotional intelligence'. There's even evidence that similar brain structures are involved. It's a similar basic challenge: understanding minds. Sun Tzu advised budding strategists not just to 'know your enemy', but also to 'know yourself'. And humans are biologically equipped to play that game.

It's still not easy, even so. Take self-awareness. Two quotes from earlier American Presidents neatly illustrate the point. Reflecting on his time in office, amidst the turmoil of Vietnam, President Lyndon Johnson claimed that, 'Presidents alone know how and why they reached particular decisions and the many considerations that affected them.'[14] Perhaps so, but his predecessor in office, John F. Kennedy offered the more astute take, observing that, 'The essence of ultimate decision remains impenetrable to the observer—often, indeed, to the decider himself.'[15] This view would find favour with generations of psychologists. We are poor at understanding why we behave the ways we do, though much better at constructing retrospective stories to explain what we did. And if understanding ourselves is tough—understanding others is harder still.

Kennedy applied himself to the challenge of empathising with his enemy. What did his Soviet counterpart Nikita Khrushchev intend to do over a series of crises, including confrontations in Berlin and Cuba? Kennedy knew that any misstep could result in nuclear war. At their face-to-face encounter in Vienna, early in his Presidency, he urged Khrushchev to avoid any miscalculations, which might lead to escalation and Armageddon. And he explicitly counselled his own subordinates on the need for empathy with the enemy. We must, Kennedy urged, give the other fellow room to back down whilst saving face. A humiliated enemy might not feel able to step back from the brink, not least because his personal rivals at home would be emboldened. This was an astute understanding of human nature, and one that eluded some bellicose advisors, especially in the military, who advocated a much tougher line.[16]

As Kennedy intuited, strategy for humans involves a delicate dance of minds. The task is to put oneself into the mind of the adversary, to understand their perspective and, critically, their feelings about the situation. That's what President Kennedy was attempting in the early 1960s. He got it wrong, in part—notably when spelling out America's lead in nuclear weaponry, hoping to sober Khrushchev up to the realities of American dominance. Rather than be deterred, a humiliated Khrushchev gambled on positioning nuclear missiles in Cuba—a highly inflammatory move. But at key moments JFK's judgment proved sound. At the height of the Missile Crisis, Kennedy was astute in judging that his rival would back down, in exchange for a covert quid-pro-quo deal to remove American missiles from Turkey. From Kennedy we learn that the dance of minds is imperfect, even for the most empathic. But as Cuba showed, reading the intentions of others is the best guide to strategy.

Machines lack empathy

Today, even the most intelligent machines have almost no capacity to intuit what others feel or think. The famous Turing test, where a machine (or rather its designers) attempts to fool a human into thinking it is also human, is a nice party trick, but nothing more. The machines, at least those that have been deployed in Turing test competitions to date, aren't engaged in theory of mind calculations like ours when we converse. If anything, they rely on our tendency to overly-anthropomorphise, given the slightest opportunity. Humans are duped into thinking the machine is human by its stock generalisations, or vaguely plausible glib answers. Sure, you can fool some of the people this way, some of the time, but not very often.

The same is true of machines that emulate human empathy. One eerie algorithm draws a baby's face on a computer screen that mirrors the emotional expression of the human looking at it.[17] Baby X seems uncannily realistic, perhaps because its designer was a costume wizard from the movies, who once made super-realistic orcs for Lord of the Rings. But there's no emotion going on behind the scenes, and no understanding of what the human's face is conveying. Even feeling emotion is utterly beyond machines, leave alone empathic emotion. Recently breathless news coverage appeared of a robot that could 'feel pain', wincing realistically in response to electrical charges applied to its skin.[18] Could 'Affecto'—the robot's name—be a breakthrough in the quest for human-like intelligence? No, just another cheap party trick exploiting our instinctive gullibility.

Without our embodied capacity for self-reflection, it's not obvious that Affecto, or any other AI today, feels anything whatsoever. Many neuroscientists argue that a degree of self-awareness is integral to experiencing pain. Self-awareness also seems critical for our understanding of other minds. Indeed, the capacity to model other minds is one plausible rationale for the creation of human consciousness. This idea of mind as a social adaptation has appealed to generations of philosophers and psychologists. Our densely connected social world demands a sophisticated self—adept at understanding what others believe and intend. Psychologists call this 'theory of mind', and humans are amazing at it, though certainly not perfect. Machines lack it entirely.

The way in which humans gauge the intentions of others is, like much else in the mind, imperfectly understood. But one distinction often made in the research is between 'cognitive empathy' and emotional, or 'affective empathy'.[19] Cognitive empathy is a more reasoned, deliberative process of thinking about others. We can think of it as perspective taking—if I were in your shoes, and knowing that we are both sensible people, what would I decide to do?

Animals can take the perspective of others too. Crows, for example, will re-hide food if they think they've been observed hiding it by another bird. This is a basic form of 'mind-shifting'. Deception like this doesn't even require a rich theory of mind—it's a behaviour that can be learned via reinforcement learning. If another crow is present, my food will get stolen. If it's not, I'm ok to leave it here. That might be considered enough for strategy at a basic level. But it's pretty crude. It's not even clear that the crow is actually adopting the perspective of the other bird, so much as responding to its presence.

A machine can definitely learn to deceive in that simple way. Like the crow, it could even learn to understand gaze—another animal is present, and it's looking my way, therefore I should move the food when it turns away. When payoff is immediate, algorithms can learn to respond to the behaviours of rival agents, including using strategies to deceive them. After all, that's just pattern recognition, and we know AI is excellent at that. *If this, then that*. In one salient bit of research by Google, two agents were seeking to communicate, a third to intercept their communications. In this particular experiment, the code-*makers* beat the code*breaker*. The two agents developed a novel approach to secure message encryption. No theory of mind here, just a reward in an adversarial, competitive environment.[20]

What about situations in which the reward or penalty from perspective-taking is not immediately apparent? It's essential for good strategy to figure out what others intend to do in the future and also what our future selves will want. That requires imagination. The key skill is to learn what others *intend* to do, not from observing their prior behaviour, but from constructing a mental model of their attitudes and beliefs.

That complex cognitive skill has been the key to our complex and cumulative culture. When we say that someone 'apes' someone else, meaning that they mimic what they do, we have it exactly right. Chimpanzees copy others in their group to learn how to 'fish' for termites using a stripped branch to agitate the nest. That permits a very primitive sort of culture—the branch, or rod, is a manufactured tool, and techniques vary from one group to another. But this aping only takes the chimpanzees so far—to really ratchet up the possibilities, they would need the sort of *intentionality* that humans possess: an understanding of what's going on in the other mind that is in part instinctive, and in part reasoned. Then they, like us, could design a better tool to achieve the goal, rather than just blindly copy. Like chimpanzees, the ability of machines to achieve this level of understanding are minimal. And this severely constrains the possibilities for machine strategists.

Machines predict in 'toy universes'

Machines have become very good at making predictions in the tightly constrained universe of board and video games. That's because these simplified worlds strip out much of the rich complexity of the real world, making decisions much easier to compute—even if the calculations required are often still far from trivial. The most complex video games have a welter of possible moves. Complex sequences of actions are needed to succeed against adversaries. Even games with far fewer possible legal moves, like chess or Go, are fiendishly complicated because of the 'combinatorial explosion' that occurs when simple moves are combined. The fastest, most powerful computers of today certainly can't 'solve' Go or chess—there are simply far too many combinations of possible moves.

But it turns out that doesn't matter. All that counts is that the machines can outperform human opponents, who have very different ways of working out what to do next. Scholars refer to the machine's 'brute force' abilities—pattern recognition, infallible memory for earlier games, unerring ability to track probabilities and payoffs as they search through millions of possible future moves. In the next few chapters, we'll see that approach square off against the human capacity for imagination and creativity. Alas for expert human gamers, brute force is a winning approach in many toy universes—and has delivered spectacular breakthroughs. Yet superhuman performance at chess still doesn't come anywhere near human performance when the rules of the game are less clearly defined—in the real world of military strategy.

Bargaining, bluffing and brinksmanship

In a 'toy universe' one can make some calculations about how an opponent will respond to a move, especially if you make some simplifying assumptions—for example, that they are 'rational', like you—both trying to win the game by working out the best moves. If you can assign some sort of numerical value to each move, you can open the way for machines to predict: they'll calculate the most likely path to victory, branch-searching ahead as far as their processing power and speed allows, or looking back through millions of games they've played before, in order to pick the optimal move. May the most powerful machine win! Even in fiendishly complicated games, this is broadly still how it works today. And, for a time, there were hopes that a similar approach might also be applied to questions of military strategy.

Thomas Schelling is one of the landmark thinkers in strategic theory. The long-lived Schelling wrote his classic works on strategy in the 1950s and 60s, an era when nuclear weapons were a new and terrifying weapon.[21] And also an era when computers and quantitative thinking seemed to offer a new, more rigorous approach to thinking about all manner of social questions, including nuclear war. One exciting branch of maths was 'game theory', which considered the way in which rational agents interacted. Pioneered by the brilliant John von Neumann, co-inventor of the modern computer, game theory looked like a good way of modelling the sorts of adversarial behaviours that took place in international relations. It was enthusiastically embraced by a group of quantitative theorists then growing in prominence, some of whom worked for the US Air Force's inhouse thinktank, the RAND Corporation.[22] Schelling himself was a mathematically trained economist, and so was well placed to take advantage of the new technique. But what makes Schelling's work of enduring importance is his use of metaphor, not maths. His strategists were emphatically men, not machines.

To pare away the complexity of the real world, game theoreticians make some bold simplifications about human minds. They typically assume, for example, that people have distinct preferences that they would like to achieve, and that these preferences are logically consistent, or 'transitive' in the jargon. That is, if I prefer option A to B and option B to C then I should prefer A to C. If: then. Already, you can see how this sort of logic might be just what a computer could readily work with. Perhaps making strategy with AI would be as simple as teaching a warbot some game theory?

The problem is that the real world rarely looks like that theory. The simplifying assumptions that game theorists sometimes employ are rather unrealistic. If that nonetheless produces meaningful results, that might be a small enough price to pay. Does it though? Economists, who eagerly embraced the approach, could plausibly assume that the many actors in a market economy and the pressures of competition between them combine to make game theory a useful way of modelling aggregate behaviours. But what about international relations, with fewer actors, and where there are many peculiarities to each crisis?

More sophisticated game theorists relax some of their simplifications, for example by introducing uncertainty about the payoffs of various options and the odds of achieving them. One interesting variation is the idea of using a 'mixed strategy', where players deliberately inject chance into their decision-making. In a normal game, a player who knows their own options and those of the enemy would theoretically be able to calculate out their best move, and vice versa—as on the chess board. Symmetrical knowledge of the game doesn't deliver inevitable success in the real world, otherwise I'd be world chess champion. But it does at least make the game susceptible to computation—the best calculator of moves wins. That plays to a machine's strengths, of course—allowing it to search deep into a game. That's precisely how machines play chess—like brilliant game theorists. In a mixed strategy, by contrast, the point is to surprise the adversary by surprising yourself—with pure randomness. Think of using a loaded dice, or a weighted coin toss: gambling, in other words.

Schelling understood that real life wasn't as clinical as the game theories suggested. His writing is rich with psychological detail—about emotion, status and chance. And Schelling knew all about mixed strategies. Consider the problem of nuclear deterrence. Clearly, this rests on a bluff. If the bluff is called, everyone loses. The essential contradiction at the heart of nuclear deterrence is the threat to obliterate yourself in retaliation to an enemy attack: Mutually Assured Destruction or MAD, as it's famously known. By the early 1960s, both sides in the Cold War possessed sufficiently large nuclear arsenals that neither could be certain of destroying all the enemy's weapons in a 'first strike' attack against its weapons. It was likely that some weapons would survive a first strike and constitute an 'assured' second strike capability. The introduction of hardened silos, mobile launchers and nuclear armed submarines all served to increase the survivability of a second strike. But how on earth could you persuade the enemy that you were serious enough about deterrence to strike back?

If it ever came to war with the Soviets, the Pentagon's plan called for all out nuclear attack on the Communist world. Their Single Integrated Operations Plan (SIOP) would use every available bomb in one almighty assault.[23] But even that would not be certain of destroying the enemy's forces. The Soviets might still have some surviving capacity to respond. Tens of millions of Americans might die, even still, as the retaliation struck home. Utter madness. No one in their right mind would contemplate unleashing such devastation, would they? It would be immoral and abhorrent, not to mention suicidal. And to do that in response to even a minor provocation was simply not credible. The great irony of nuclear strategy was that unless you could persuade the enemy that you were serious about retaliating, including with nuclear weapons, the entire edifice of deterrence would collapse on itself.

But how? The clue for Schelling was 'in their right mind'. You would have to persuade the enemy that you weren't quite in control of your own behaviour. His solution was to inject a smidgeon of randomness into your decision-making. A 'mixed strategy', in other words. Ever imaginative, Schelling deployed a metaphor to explain his thinking. He imagined two men, tied together near the edge of a cliff. If one went over, to his death, the other would too. It was a compelling analogy for MAD. How to bargain with the other man? Schelling suggested dancing erratically on the slope leading to the cliff edge. At some point, you might lose control of your ability to pull back from the brink. Or you might not—it would be hard to know when the point of no return had been reached. Hopefully, that degree of randomness would persuade your antagonist to back down while you still remained in control.

Later, another analogy suggested itself to Schelling. It was the opposite of a mixed strategy—taking away all the uncertainty. This time he imagined war as a game of 'chicken' between two young men, racing their hot-rods towards each other on a long, straight highway. Crowds of cheering supporters look on, heightening the emotion and the stakes. Who will back down first and swerve their car out of the way? Lose face, or lose your life? For Schelling, the thing to do is to remove the option of backing down. He suggested that one driver unscrew the steering wheel, hold it aloft, so that the other driver could see it, and then throw it from the car. Now the choice was, literally, out of his hands. There was, however, one big problem—what if the other driver had the same bright idea? Mutually assured destruction would ensue. Both strategies, 'dancing' and 'hot-rods', were about 'mind reading'—getting inside the head of the adversary. You could model them formally, via game theory, using probabilities and payoffs. But the big question with both metaphors was how to translate them into international politics. Was there a real-world version of chicken?

Some genuine crises did have elements of Schelling's logic to them, whether or not that was intentional. In the Cuban Missile Crisis, for example, there was plenty of randomness to contend with, like the accidental U2 surveillance flight that strayed into Soviet airspace in the middle of the drama. And sometimes leaders deliberately cultivated the erratic, even unhinged, behaviour of Schelling's man on a clifftop. President Nixon became the arch exponent of what he termed the 'madman theory' of statecraft—crafting an image of an irascible leader who might take dangerous escalatory steps unless the enemy backed down. More recently, President Trump has also portrayed himself as a foreign policy maverick on a short fuse.

But when the really big crises came JFK's way, in Berlin and Cuba most notably, another very human feature emerged—crippling fear and extreme risk aversion. Game theorists, ever helpful, had a term for this too—a 'minimax regret' strategy. This was an approach that sought to minimise the maximum loss that might arise from any decision. The maximum regret induced by unleashing Armageddon is considerable, and leaders have, thankfully, chosen to step away from the brink, rather than goad enemies over the cliff edge. The prospect of nuclear annihilation has acted as a sobering brake on gambling.[24] At the brink, in October 1962, that's exactly what Khrushchev and Kennedy chose to do. The awful prospect of nuclear war chilled the deliberations of the antagonists. Indeed, a minimax preference has reinforced the nuclear taboo over the decades since Hiroshima. And perhaps we shouldn't be too surprised—humans are, broadly, loss averse, and nuclear bombs amount to a catastrophic loss.

We can learn much from the big international crises of the Cold War years like Cuba; not least that coercion to compel and deter enemies is an uncertain affair. Using force is not rational in some abstract, game theoretical sense, in which probabilities and utilities are known and carefully weighed ahead of any action. We also learn that pure chance sometimes plays an outsize role in determining what happens. And perhaps most importantly, we can learn that emotion informs human judgment. Fear of the devastation in a general war exerted a powerful brake on escalation. Anger at being humiliated by Kennedy led Khrushchev to position missiles on Cuba, and then anger at being blindsided led Kennedy to take a firm line. Fear kept both men from the brink.[25] In the end, as the ever-astute Clausewitz wrote, 'action can never be based on anything firmer than instinct, a sensing of the truth'.[26]

War games

And intelligent machines? How would they go about 'sensing the truth' in the real world? Would they too be risk averse when it comes to nuclear escalation? One thing is abundantly clear—they would do things very differently from Kennedy; by playing the odds, like game theorists, not mind reading like world leaders. And if they injected some randomness into proceedings, then unlike Schelling's man dancing on the clifftop, they wouldn't be remotely interested in observing the whites of the other person's eyes—just on calculating probabilities. If you do this, I do that.

If i-War ever becomes a reality, warbots will decide on escalation. That was, after all, one of the options for DARPA's Deep Green project—the ability not just to shape operational decisions by humans, but to make them itself. We'll need to think carefully how they go about it. Perhaps now, armed with some insights from human strategy, we're better placed to understand how intelligent machines might shape warfare.

At the most tactical level, in battle, the mechanics of warbot escalation might not matter all that much, at least beyond those involved in the fight itself. A small armed drone comes under fire, and communicates to the swarm operating nearby, which diverts other drones to suppress its attackers. This is probably what most people have in their minds now, when they think about 'killer robots'—an armed Predator drone. But the pressure of combat could push this sort of autonomy ever upwards. Smaller actions—'contacts'—are knitted into larger engagements; engagements become battles; the battles fought by these tactical units need sequencing. Is this the moment when a human commander can come in 'on the loop' to weigh further escalation, and to be legally and ethically accountable for the consequences of doing so? Perhaps. But if the enemy is content to leave that sort of decision to a more senior commander, the human commander might be defeated by speed, by the machine's more astute tactical judgment, or by some other advantage of tactical autonomy.

Somewhere between an autonomous tactical drone armed with a shotgun and thermonuclear war there must be a firebreak where human judgment can be brought to bear. Time is critical the closer you are to the action. As we move to the more elevated levels of command there should be more time for human judgment. The problem is that it's not abundantly clear where that level is, or what factors will shape it. There are huge risks here. Some degree of autonomy and AI is already incorporated in weapons systems that are capable of extremely high levels of violence. At the moment these are predominantly defensive systems, and the autonomy is there largely because the system needs to be inhumanly fast. But that need not always be the case, and the line between offensive and defensive weapons is rather subjective.

And there's a further risk: the sorts of weapons that can deliver nuclear weapons are abundantly capable of being deployed with non-nuclear warheads. Until recently the ballistic missile was the most assured way of delivering a nuclear warhead. Incredibly fast, and traversing the upper reaches of the atmosphere, the intercontinental ballistic missile was impossible to defend against. It still probably is, at least when combined with multiple re-entry vehicles that can saturate any available missile shield. But it's perfectly possible to combine nuclear warheads with cruise missiles, torpedoes and—currently in development—hypersonic missiles. A defensive system may be unable to determine how powerful the payload of any attacker is. In which case how can we limit automated escalation by instructing our machines to escalate only up to the point of the nuclear threshold. We had better figure out machine escalation! But how?

For the beginnings of an answer, we can look again at what happens when machines play actual games. Games are far simpler than real world strategic crises. Even the most complicated wargames strip out plenty of detail. We might not learn everything about strategy in a nuclear crisis from computer simulations and video game play. But that doesn't make them irrelevant—they can still provide a good test-bed for thinking about machine coercion in war, in much the same way that the RAND strategists employed game theory. Perhaps they're even more useful: Game theory doesn't adequately capture the emotional aspects of human strategy—especially the terrifying fear of nuclear war; but that's lacking for machines, who might go about real-world brinksmanship in much the same way they do in a game.

Where better to start than with poker—the game that Clausewitz surely had in mind when comparing cards to war, in the epigraph of this book. In recent years, poker has joined chess, Go and Atari in succumbing to the deep learning revolution. This is a remarkable breakthrough, because unlike chess and Go, poker is a game of imperfect knowledge. In 2017, Libratus, a poker playing system from Carnegie Mellon University managed to beat world class human opponents, winning $1.7m in chips from four top ranked professionals.[27] Game theoreticians would not be surprised by some of the techniques it employed, including 'counterfactual regret minimization'. There are echoes of 'minimax' here—the computer minimised its regret in similar fashion to the way that Kennedy and Khrushchev strove to reduce the anticipated regret they might feel after blowing up the planet. In a nutshell, the computer played itself over and over again at poker, many millions of times, and for each decision it made, it later looked back to determine how another choice would have played out versus the one it actually made. By minimising the gap between its move and a better alternative, its algorithm would eventually converge on the best possible one. This is known to game theoreticians as the Nash equilibrium, after the brilliant mathematician whose story is told in the movie *A Beautiful Mind*. At the Nash equilibrium, there's no alternative that can deliver a more effective decision against an opponent similarly minded. Of course, the element of luck in poker means there is no definitive, objectively optimal choice—just like choices in the real world. But all that's needed is the closest approximation. That's especially true because any humans Libratus plays will fall a long way short of the Nash equilibrium, even world class ones. Libratus enjoys a formidable memory for its earlier hands and it has the capacity to look ahead far into the game, working out the 'what ifs', even in the impossibly large space of possible futures in a poker game. Those are strengths of modern AI—incredibly swift calculation using massively powerful processors and an unerringly accurate memory.

There's more. Libratus also used its formidable powers of pattern recognition to find and exploit regularities in the style of its human opponent—the extent to which they depart from that Nash equilibrium. It tweaks its approach as the game unfolds and it learns the opponent's style. Unlike a machine's random number generator, humans really struggle to come up with truly random choices. You might, for example, theoretically know that a string of consecutive numbers in a lottery is just as likely as a wider spread, but it certainly doesn't *feel* as likely—it looks ordered, not random. So rather than look for a physical 'tell', like James Bond in *Casino Royale*, Libratus instead noted and remembered glitches in the way particular players played particular cards. By contrast, the human adversary, perhaps used to discerning 'tells' in other players, would have no way of getting a handle on Libratus's thoroughly randomised style. After six days of poker, the pots were about even, and the world class poker stars were holding their own against the machine. But thereafter raw computing power allied to machine learning kicked in and the humans were soundly beaten. After twenty days of gameplay and some 120,000 hands of poker, the results were clear—poker had gone the same way as Go and chess.[28] And now, just a few years on, the system has been further refined. Libratus took on players one at a time. But Pluribus, the remarkable state-of-the-art poker player from Carnegie Mellon, can currently beat five human opponents in a six-player game.[29]

If you play online poker, Libratus should be food for thought. But what does it mean for machine strategy? It's clearly been food for thought for some in the Pentagon. The team behind the poker engine has since brought its approach to defence, with their start-up company Strategy Robot awarded a multi-million dollar contract from the Defence Innovation Unit, a branch of the Pentagon tasked with accelerating the adoption of cutting-edge Silicon Valley technologies. Separately, DARPA is soliciting for computational game theoretic approaches to military decision-making. Their project envisages applications that can handle 'multiple interacting agents, extremely large search spaces, sequential revelation of information, use of deception, continuous resource quantities, stochastic outcomes, and the ability to learn from past iterations'.[30]

It sounds like a strategic AI. That's the dream at DARPA anyway—the current state of the art remains a long way from that. One rather large problem that the budding AI strategists will need to factor in is that, unlike Libratus, they probably won't get 50 or more hands against the adversary to improve its gameplay. 'Solving' real strategy problems will require much more than solving poker. In the real world, decisions are altogether more complex than card games, whatever Clausewitz wrote. Something else is needed—but what?

THE ART OF i-WAR

Libratus, the poker playing AI, isn't a genius. The geniuses are the mathematicians who built it—John von Neumann, originator of both the computer and game theory; John Nash, discoverer of the optimum strategy for two-player games; and the Libratus team at Carnegie Mellon University, who harnessed some abstract maths to a powerful deep learning architecture. To see the difference and appreciate why it will be difficult to find a genius warbot general, we need to think about creativity in humans and machines.

Libratus *is* somewhat creative, but only in a very limited sense, and one far away from what we humans are capable of. There's an important distinction that will constrain the ability of AI to contribute to strategy. World chess champion Magnus Carlsen is a creative genius, possessed of preternatural intuition. Exploring the basis of that genius will help us understand the profound challenges facing DARPA in its quest for strategic AI. Clausewitz, as we saw, thought military commanders should also possess what he called 'genius'. This, in the philosophical argot of the day, was someone who could bend fate to his will, seizing opportunities that mere mortals could not even see. Genius required the ability to make decisions on the basis of imperfect information, in times of great stress, without being swept away by emotion. Perhaps above all, it required more than a mechanistic following of rules or precepts. Though Clausewitz didn't stress it, his ideal commander was creative.

Creativity and genius continue to fascinate psychologists and the general public alike. The popular science writer Malcolm Gladwell argued provocatively that practising something for 10,000 hours would be enough for anyone to attain expert level.[1] But plenty of people *do* spend that many hours honing their skills at some pursuit or other without ever achieving more than rank mediocrity—whether that's learning a language, struggling with maths, or spending a lifetime bashing out banal watercolours or clichéd poetry. 10,000 hours of practice might be enough to groove whatever the activity is in some way—rendering conscious deliberative actions into highly automated and subconscious ones. But that still doesn't make for genius. Libratus certainly practised diligently, playing itself many thousands of times. That's the essence of deep learning, after all. But like the devoted yet mediocre artist, it's not always enough.

Practice certainly makes things easier—partly by making them automatic. The golfer who 'overthinks' his shot while in a poor run of form is likely to compound his errors. Instead, psychologists talk about 'flow'—the ability almost to dissociate our deliberative actions from the unconscious—to slip into a well-worn subroutine. Someone who has learned the complexities of riding a bike (you never forget) is invariably totally flummoxed by a bike where the steering is reversed, so that turning the handlebars left steers to the right. Give it enough time, though, and a new cognitive subroutine develops that masters the complex forces operating in reverse. Now riding a normal bike will be just as challenging, until with a sudden shift of cognitive subroutine the old neural pathways are reactivated and the original bike can be ridden again.[2]

The brain of the cyclist is on autopilot—as it is for much of what we have learned to do. After all, that's one way to think about learning: the strengthening of associations between neurons, so that networks fire more readily. This is what allows us to plan our days—not by working out every last detail and performing some cost-benefit analysis of what we should do and when. Economists might think that's how humans behave, but psychologists know better. Instead we are creatures of habit—we've learned to group certain actions together to produce good results. And we use these patterns, or scripts of behaviour to get through the day, preserving scarce cognitive resources for when we really need them. Psychologists sometimes refer to 'chunking'—the bracketing together of individual bits of intelligence that work well together. If one item is in play, the others are too, or at least they might come more readily to mind.

And so, practice helps raise skill levels and groove routines; it gives us scripts and chunks of information that we can roll out on autopilot. It gives Libratus and Arnold Palmer alike the honed performance that allows them to excel. But does it help with the creativity needed for genius, including genius in war? We can all learn the rules of chess, and practise for many hours so that gameplay becomes natural, even automatic. With enough effort, we can study games, and remember classic openings, allowing serious players to move into a game without really engaging much cognitive effort in mapping ahead. But we aren't all Magnus Carlsen, or Napoleon. Creativity seemingly requires more than practice. But what?

Three types of creativity

Margaret Boden, whose research spans philosophy of mind and Artificial Intelligence, distinguishes three forms of creativity, which are as useful in thinking about chess grandmasters and jazz greats as they are brilliant military commanders.[3] They are also a way of thinking about machine creativity too, which is precisely what Boden does with her typology, as we shall here.

For Boden, the signature of creativity is a behaviour that is at once novel, surprising and useful. Her first form of creativity is pretty straightforward: it involves new combinations of ideas. For jazz musicians and chess champions, creativity is about making something new by combining elements in new ways. But the new combination can't just be nonsense, or random noise; to be creative it ought to serve some purpose. That doesn't necessarily mean winning a game, still less a war. The product of creativity need not be *useful* in a narrow utilitarian sense in which some problem or other is solved. But it does mean being grounded in some context in which it has meaning.

So, jazz improvisation has to work within the context of music and, specifically, the conventions of jazz music. It will involve melody, timbre, rhythm, harmonics. It will utilise the 12 tones and semi-tones of the musical scale. The notes will be pitched within the hearing range of the human ear, not least because the jazz pianist will be playing a piano. And yet, despite these conventions of the genre, there is plenty of scope for creativity. 'Said and Done' a piece by the brilliant modern composer Nils Frahm starts with a single note played over and over again, for minutes. It's hypnotically repetitive, but there's variation too—in the loudness of the notes and their attack. And then, just as you are focused on the Zen-like repetition of that single note, it blossoms into richly textured harmonies. Creativity is found in new combinations, provided they are grounded in context—meeting our expectations of the genre in some respects, violating them in others. When Frahm and his collaborator Ólafur Arnalds improvise together on the piano, they are assembling combinations of notes collectively in the moment, anticipating where the other player is going, and working within the patterns already established to create something entirely new. Modern language translation in AI does something similar, as we'll see.

What about the chess genius? Is combinatorial creativity useful there? Certainly, but not in the way you might immediately think—of combining sequences of moves. Like many chess masters, the former world champion Garry Kasparov kept and studied voluminous notebooks of famous encounters, including his own. Then, when computers became ubiquitous, he could consult even more games, calling them up at the touch of a button—a form of extended or augmented cognition that we'll return to.

Kasparov chunks moves. When he plays a game, he doesn't spend the opening exchanges like an amateur, weighing up the pros and cons of each individual move.[4] That would be mentally draining and would concentrate his attention disproportionately on the tactics of the board—how the near-term moves will play out—rather than the strategy. In fact, that laborious, iterative approach is exactly how chess computers play the game—they are just rather better than us humble amateurs at working through the possible combinations of moves, using their computer processing strength and formidable memory. We don't have to go far before we blunder—'whoops, I didn't see your knight there!'

Kasparov, meanwhile, can navigate the early part of the game on autopilot, because he's seen many of the patterns before. The positions are 'chunked' together in his mind and make sense at some sort of aggregate level. Brain imaging studies of expert players suggest that visual and spatial reasoning might be involved in seeing the group as a whole, rather than in terms of individual pieces. And chunking like this allows him to focus on developing the game *strategically*—concentrating power in some part of the board, developing a position for a few moves down the line; sacrificing a piece in order to buy time and space elsewhere on the board; or just playing a holding pattern to see how the opponent seems to be developing their own position. But then, in the mid-game, out beyond the pre-ordained, rehearsed playbook of the well-prepped grandmaster comes the scope for creativity; for new packages of moves that can surprise or unsettle the opponent.

In chess, the potential for combinatorial creativity seems much more limited than for an improvising pianist. Creativity is restricted by the 'toy universe' of the chess board. But there's some scope for originality—at a more elevated level than the individual moves. The real combinatorial creativity comes from the combination of the strategic positional play, with its chunked information, and the iterative tactical looking ahead—searching out the most profitable moves in any moment. In the mid-game, both mental systems are involved in the mind of the grandmaster, innovating and improvising amid bewildering complexity. The scope for combinatorial creativity in Frahm and Arnalds' studio is almost infinitely greater. And in the real world, where objectives are far broader still than creating aesthetically pleasing music, the scope for combinatorial creativity is essentially limitless. That's the unbounded world of war and strategy.

AI does exploratory creativity

Boden's second form of creativity is 'exploratory creativity'—the ability to look deep into a given domain and discover what hasn't been found before. Machines excel here. This is the sort of creativity employed to devastating effect by the chess computer Deep Blue that dethroned Kasparov as world champion in the 1990s. And it's the sort employed by Libratus in upending the world of poker. Libratus is orders of magnitude more powerful than Deep Blue, and its mathematical techniques are more refined, but it's really doing the same thing. AI searches ahead at lightning speed to weigh the payoffs from moves. The space in which it searches could be huge—as with Go or poker. Even the fastest computers can't search all that. But all that's needed is that their capacity to find useful, new information in the search space will allow them to outperform the combinatorial creativity of their human adversary. In the bounded world of the chess and Go boards, that's possible. When it comes to warfare though, it's another matter entirely.

In chess, a computer can search deep into a game, but the *breadth* of its exploratory creativity is limited. Chess is a tightly rule-bound game in which you can't suddenly discover a new move. There *have* been changes in the rules of chess, with variation in local practice, for example over how the Queen moves, or how much time you have to take your turn. But these are well regulated and usually known to both players: you cannot grasp a sudden advantage by doing something radical, like moving your rook along a diagonal. You can, though, explore the possible moves ahead as thoroughly as possible, which is what a chess computer does. The computer can't 'solve' chess this way—rather it focuses on the most likely moves, discounting the others in order to explore more deeply the pathways that seem more profitable. And even here with the rise of supercomputers and the defeat of the human chess champion as far back as 1996 there is some scope for novelty; the boundaries of exploration have not been reached.

In the board game Go, exploratory creativity has also proved important in the struggle of man against machine. After chess, Go became the yardstick for AI gameplay, because the size of the search space dwarfs that in chess. In Go, as is often said, there are more possible moves than atoms in the universe. Surely, observers thought, the raw computer power required to search deeply into that domain would be beyond modern AI? Not so. In 2016, DeepMind's AlphaGo demonstrated that computation trumps human skill, even in a truly vast search space. In deposing Lee Sedol, the world champion, AlphaGo's limited version of creativity overwhelmed the richer human variety.[5] Go was a formidable computer science problem, but still just a board game. Nonetheless, when all was done and Sedol defeated, the match clarified some big things that have relevance for our study of AI strategists.

Chief among these was confirmation that if AI is creative, its creativity differs from what we usually mean by it. At move 37 in game 2, the computer stunned onlookers and Sedol by making a radical move, one vanishingly unlikely to have been played by an expert human. There were gasps from the commentators and a startled Sedol left the table to ponder his reply. It was a game-winning move, which aficionados eagerly attributed to AlphaGo's ingenuity, or creativity. But was the machine really 'inspired', or were the onlookers just anthropomorphising? In Boden's terms, the move was new, surprising and valuable—and so genuinely creative. But it was simply exploratory creativity on steroids—searching in the available universe of possible moves for a new angle that would bring a marginally better probability of success, many moves further on. Because humans, even expert ones, don't search so extensively in pursuit of marginal gains, the move looked highly novel. Sedol was stunned too, returning to the table to think for many minutes, before going down rapidly to defeat.

Yet something was lacking. If it *was* creative, AlphaGo certainly wasn't creative in the sense that Sedol was. Like other genius players, Sedol didn't see the board as an exercise in number crunching, but as a combination of visionary strategic play and short-range tactics. And there was an intensely psychological dimension to his approach. Sedol described looking across the board, as you would in playing a human. But the human on the other side was only there to move the pieces for the machine. It was, Sedol reflected later, like looking into a mirror, staring into your own soul: the doubts and uncertainty he found there were crushing. Bravely, Sedol rallied in a later game—finding that one way to match AlphaGo's unexpected move was to play in an unexpected fashion himself. He did so, and the machine was duly defeated, its exploratory creativity momentarily unbalanced by Sedol's combinatorial creativity. Game theoreticians and readers of Schelling would be proud of Sedol's approach—it looked something like a mixed strategy, or at least one that forced the machine off its thoroughly computed statistical pathway to victory.

Exploratory creativity is the singular advantage of computer intelligence over human—resting on the ability to rapidly explore masses of data, to calculate probabilities and to faithfully remember what it has learned. In the closed universe of a board game, even with trillions of possible paths down which the game might progress, the computer can explore far more adeptly than the most skilled human. AI is a statistical processor par excellence, and that affords tremendous exploratory strength.

Not so combinatorial creativity. We could argue that AlphaGo demonstrates combinatorial creativity, in the sense of combining sequences of moves in new ways, but that doesn't meet the definition. The combinations of moves the computer makes is ultimately just the result of its deep exploration, working backwards, which is highly tactical. To demonstrate combinatorial creativity within the game's severe structural constraints, a player could combine strategic thought at some macro-level while still being able to engage in tactics, move by move. That's the genius of Sedol and Kasparov—and it's the skill of human strategists too.

Not long after his defeat, Sedol retired from professional play, crushed by the experience of losing to an opponent that, as he put it, 'cannot be defeated'. But designers of military AI shouldn't get too excited. Attempts at developing a strategic AI along the lines of AlphaGo face some formidable problems. It's hard enough to create a battlespace manager of the sort fielded by DARPA in the 1980s. The options there are more constrained than at the strategic level, more so for being at sea. But they're still orders of magnitudes more complex than Go or chess, and both those games end up with a vast search space to explore. As for DARPA's Deep Green project of the early 2000s, the ostensible goal was to assist decision-making at the operational level—knitting together missions by brigades in all sorts of conflict environments. The idea was that Deep Green would simulate the clash of various units, and suggest optimal plans for military staffs, allowing them to narrow their own options. Its 'crystal ball' function was designed to look ahead and predict the likely outcome of an encounter. Good luck! The project's name itself suggests the debt owed to Deep Blue, rather more than to the deep learning revolution that was only then getting underway. But even with the formidable developments in the decade since, the gains made by state-of-the-art in AI remain firmly in exploratory not combinatorial creativity. And the further away from the mechanistic clash of tactical forces you get, the less useful that sort of creativity is.

Transformational creativity is real genius

There's a final form of creativity that machines lack, but that Boden highlights. This is transformational creativity—the ability to transcend the search space entirely and open up a wholly new vista of possibilities. It's this aspect of creativity that ensures strategy will retain its indelibly human element for a long time yet, even as machines become more powerful at search and combinatorial creativity. Transformational creativity in war goes far beyond the battlefield clash of warbots, even in combined arms formations, with their myriad combinations of military units and weapon systems. It extends strategy into more ephemeral and existential territory that is richly psychological. Why do we fight? What do we want, and how badly? Good luck computing that.

Transformation demands the ability to operate in an entirely new milieu. And AI is rather poor at adapting from one task to another. This, though, is starting to change—MuZero, the latest iteration of DeepMind's all conquering Go algorithm has learned expert level play in Go, Chess, Shogi and all 57 Atari console games, all without any prior knowledge of the rules.[6] Is that really transformative creativity? It would be if I did it, because I would be (attempting) to develop some sort of tailored approach to each game—the usual human blend of tactical ability with strategic thinking that we saw from Sedol. Different strategies would be required for each game, but for the machine, the same essential process is at work—narrow the search field, and conduct reinforced learning many, many, many times, so that some probabilities can be associated with each move. Use these probabilities to work out the best course of action, and voila!—you've computed your way to victory. All else being equal, the machine with the largest amount of computer power will win. It will be able to conduct more training games in parallel, and to peer deeper into the game than its adversary. In other words, MuZero demonstrates exploratory creativity, masquerading as transformational creativity.

We humans go about things differently, although precisely *how* remains unsettled. At some level of abstraction, there are general cognitive skills that might be useful in separate and very different domains. Being creative in the transformational sense could involve recombining this basic package of skills in a novel way that is useful in the new context. For example, breathing control learned for swimming might be repurposed for wind instruments, and fine motor control that is useful for writing might help with strings—and vice versa. Some of these core skills might be innate, or hardwired by evolution—like the way we make sense of vision by innately perceiving things as lit from above—because that's where the sun invariably is. Or consider our 'mentalising' or mind-reading skills—our empathy is often instinctive, not reasoned. Other skills are learned from practice, as with the many years spent irritating the neighbours while a child screeches her way through Beethoven on a violin. Most skills probably involve some combination of innate and learned ingredients—as when reading borrows the edge detection systems we've evolved in our visual cortex.

We might conceivably make machines along similar lines—and arch connectionists like Yoshua Bengio think that's feasible. Could not a network of specialised AIs with pre-tuned nets replicate our innate, basic skills, and then get creative by repackaging them? Currently, DeepMind is trying something along those lines with a neural network that figures out causal relationships between visual objects.[7] That's something toddlers do well. If it can grasp that essential cognitive building block, perhaps the AI could generalise from it, combining it with other core skills to create something remarkable.

The difficulty is with the repackaging. Creativity isn't just about mixing things up—it has to mean something. Otherwise it's like expecting chimpanzees to type Shakespeare, or hoping that if you throw enough mud at the wall, art will somehow emerge. But that's not how the incredible prehistoric horses appeared on the wall of the Chauvet cave in France, some 30,000 years ago.

All the ingredients in our skill toolbox arise from and are in the service of our bodies. Our emotions, our senses, our memory, our spatial and temporal reasoning; all are anchored in our embodied experience. Narrowly, of course, those bodies want to survive and reproduce. At some fundamental level, that might look like the sort of reinforcement learning problem a deep neural network could grasp. But it's not. Our goals are richer and infinitely more subtle than maximising the score in Space Invaders. They're informed by the context of our intensely social world, and our place within it. This makes strategy, as pursued by humans, a richly creative process, akin to art. Yes, on one level strategy serves a purpose, to secure us against threats, but that's only the beginning of strategy, not the end. Security depends on our group; and our group is a shared cultural context, rich in meaning.

AI doesn't understand, so can't create

DARPA knows all this. That's why the goal of their AI Next programme is computers that 'understand and reason in context' and that function 'more as colleagues than as tools'.[8] It's vaultingly ambitious, and almost certainly doomed to fail, at least with computers as we understand them now. Why?

To put the various skills together creatively, we have to want to put the effort in; and to understand what needs to be done, at some level deeper than the immediate gratification of value given by a reward like the score in a game. We are highly motivated by the search for meaning. We are, in a word, curious. In the 1940s, the psychologist Abraham Maslow proposed a 'hierarchy' of human needs, in a pyramid shaped diagram that has achieved enduring fame among generations of social science students.[9] At the base of his pyramid were physiological needs, like food and water, sleep, shelter and so on. At the apex, the more intangible need for 'self-actualisation', which Maslow suggested could come from achieving one's potential, including through creative activities. Play is an important aspect of that creative process and, related to that, imagination another, whether in daydreams or even in our sleep. There we find space to explore, re-combine and transcend what has gone before—safe from physical repercussions and even from social judgment.

For all the human ingenuity that has gone into making it, deep learning AI is still in essence a wonderfully sophisticated abacus. This severely limits its creative abilities—especially that which transcends the norms of whatever task it's performing.

The abacus, no matter how refined, lacks curiosity, playfulness and other human motivations besides. These are all about exploring, but not in the sense of mapping a domain space, like AlphaGo—rather, in the sense of finding a new game altogether. They are part of our search for meaning.

At root, what the AI and the human *want* is very different. Over the chess board, both can be said to want to win—but for the AI that's about maximising some numerical score—optimising a reward function. For the human, however, there's much more involved. The game can certainly bring material rewards; it may even bring status and esteem. But there is also the existential dimension of play. That's why Sedol retired from Go after his crushing defeat by a computer. Where was the meaning in being beaten by a souped-up abacus? Where was the artistry? The machine didn't *want* to compete. If it could be said to want anything, that was only to maximise a probability, or optimise some function. And if you could be defeated by something so singularly lacking in creativity, what was the point in playing at all? Go went in an instant from being an exercise in human creativity to being a sub-field of mathematics.

This distinction is hugely important for strategy. Machines that optimise against a score will have a narrow conception of strategy, as well as limited creative capacity to succeed in it. In battle, warbots wheel in aerial combat, seeking to out-compute their adversary and out-cycle their OODA loop. It's enough here to have exploratory creativity—especially if it's relentless and blisteringly quick. In strategy, more slippery concepts are at work—what does it mean to win? For humans it often involves more than surviving. That should limit our expectations of a WOPR-like system, sentient and able to reflect ruefully on the grimness of nuclear war.

If we see mental activities like play, curiosity and imagination as likewise being functions of the human mind, then the limitations of AI in this respect are obvious. Other animals demonstrate some of these capabilities too. Crows, primates, dogs, dolphins and others all impress humans with their capacity for human-like behaviours. But these are all biological brains. The challenge remains: can you have a *mind* without biology? It may not be enough, on this view, to build ever-more sophisticated connectionist AI. In the end, if the AI 'brain' lacks that biological essence, there's no guarantee that a mind like ours will emerge. So what if, instead, we bring some biology into the artificial?

HUMAN-MACHINE TEAMS

Warbots will make incredible combatants, but limited strategists. Ability at chess is definitely not the same as a genius for war. Still, even if that were all there is to it, the changes to warfare will be profound. Warbots would call for new tactics and organisations and alter the relationship between societies and armed forces.

But it's not all there is to it. The epic showdown between Sedol and AlphaGo pointed the way to a new development, one with radical implications for warfare. It's one that takes us right to the edge of scientific research today, where AI meets biotechnology. The match showed the possibilities for collaborative creativity between humans and machines. If machine minds can't be fully creative on their own, they certainly can be in combination with human minds.

Sedol retired soon after his defeat, but other humans play on. Not only are they using machine adversaries to sharpen their play; their tactics are also changing in response to the innovative gameplay of the machine. The same happened when chess computers became powerful during the 1990s. Undeterred by his defeat by Deep Blue, Garry Kasparov became a pioneer in 'centaur chess' competitions, where players would cooperate with the machine *during* gameplay. The result was a powerful combination of the strategic vision of the human player and the relentless tactical prowess of the machine.

This human-machine combination is an idea much touted in military circles. In part that's because of the ethical considerations of having machines make decisions about killing, in part because the organisational culture of militaries is built around humans. The dominant culture of air forces is that of the pilot. And for armies, much is made of the physical and moral courage needed by those who engage in close quarters combat. But beyond ethics and the organisational proclivities of these large and sometimes conservative organisations, there's a powerful rationale for human-machine teaming: the two very different forms of intelligence can combine to produce genuinely new and useful insights, covering for each other's respective weaknesses.

The centaur combination of human and machine applies more broadly than board games. Already it's shown its effectiveness in other toy universes—as with DeepMind's Quake III competition in 2019.[1] Quake is a complex strategy game in which players can team up. There are many different types of avatar with varied attributes and while tactical skill is important, so too is the ability to plan over the longer-term. As you might expect, DeepMind's algorithm reached a high-level performance, and its team of two AI players were soon out-gunning teams of expert human players. But the standout performance came from a team made of one human and one algorithm. Why? Success apparently rested on the combination of unerring tactical skill of the machine—its speed of decision and accuracy of shooting—in harness with the strategic outlook of the human, able to take a longer-range view of the game.

Meanwhile, another leading AI research firm, OpenAI, has been using another multiplayer strategy game as its testbed—Dota 2.[2] This too is a multiplayer game, pitting two teams against each other, each with five members. The game demands the familiar mix of tactical skill and strategic thinking. Here again, AI has become competitive against world-class human teams, in part on the basis of what one player called its 'hydraulics'. Once again, it's the tactical acuity of the AI that draws most attention. The team of machines never tires, or makes tactical blunders, and it's unerringly accurate. As in toy universes elsewhere, there's that now familiar story—throw ever greater computer power at the problem, and eventually that outperforms human creativity. Recently, though, OpenAI has begun to work on mixed-AI and human teams. Without any adjustment of the machine's training regimen, the inclusion of humans in the team worked well. One player, displaying the commonplace human tendency to anthropomorphise, noted that the machine 'tried to help me, thinking, "I'm sure she knows what she's doing" and then obviously I didn't. But you know, he believed in me.'[3] As if! A tournament is imminent, and the smart money would be on the mix of human strategy and machine tactics making for a powerful adversary.

Beyond competitive games, there's scope for AI-human collaboration in creating art. On its own, the combinatorial creativity of AI can produce some wild and interesting results—whether in painting, music or literature. But like the 'artwork' of a chimpanzee or a small child, it's not clear what it all *means*. For example, one interesting technique used in AI painting is the Generative Adversarial Network. Here, two algorithms work competitively to produce a picture. One AI is tasked with creating a picture that looks something like the thousands of paintings it has been trained on via deep learning. The other algorithm tries to discriminate between these efforts and the real thing. The result of one GAN artist is an eerie collage of faces that are quite otherworldly. One of its paintings, a portrait of a fictional aristocrat, sold for more than $400,000.[4]

There's some novelty value to this, but it's surely not really creative in the human sense. That would require some wider contextual meaning, within which the artist produces novel, and perhaps (for a genius like van Gogh) even transformative work. Still, humans are easily fooled. The 'Expressionist' artworks of another adversarial bot were mistaken for paintings by humans 85 percent of the time.[5] Humans were asked to rate the machine's artwork on whether it was inspiring, appealing, created with intent, and had a meaning. Clearly, they thought all of the above. But the medium of painting plays to the AI's strength—in modern art anyway there's often much sceptical discussion of what an artist means. The same is true in other art forms, like poetry. In the right circumstances AI's can get away without understanding.

But centaur approaches mean this not where our story ends. Today, humans and machines can combine to create new, meaningful art. OpenAI's GPT series and Google's BERT are state-of-the-art natural language processors. These algorithms examine words in the context in which they appear. Some words in the training passage are masked, and the algorithm tries to predict what they are. Any discrepancy between the actual word and the prediction can be used to tune the neural network. Then, given new text, the algorithm is tasked with predicting the next word, sentence, or even paragraph, on the basis of what it's learned so far. As elsewhere, the powerful combination of huge datasets and equally vast computer power generates results that are astonishing. A *New Yorker* article used GPT-2 to generate the final paragraph in each section, based on what had come earlier in the piece.[6] Here's an example. The human author writes:

> The creepy thing was that the machine was more thoughtful than I was.

And the machine adds:

> By that I mean, it seemed to want to distinguish my feelings from my thoughts. To put it another way, Smart Compose seemed to want to know me.

The human author felt the prickly sensation of the 'uncanny valley' a term that describes the sensation when a machine produces something that is almost, but not quite, human. And it's a form of human-machine collaboration: you can try it yourself online on one of the webpages offering GPT-2 text generation.

Often though, the text that the machine generates is not as slick as the sentences above. Even with the state-of-the-art version, the uncanny valley can seem impassably broad. Androids don't dream of electric sheep just yet. But the *New Yorker* piece hints at the way ahead; a future in which AI and machine can combine to create richer, truly meaningful sorts of artistic creativity—including the transformative AI that Boden describes.

The choreographer Wayne McGregor uploaded thousands of hours of dance footage to an algorithm. Trained on his style, the machine makes suggestions about possible moves in real time, while watching dancers at work. As McGregor noted, he and the machine were engaged in a 'conversation around what is choreography, who has to make choreography, what are the potentials of choreography'.[7] And there's music too. The Icelandic composer Ólafur Arnalds, who we encountered earlier improvising with his friend Nils Frahm, also experiments with human-machine teaming. On his album *Re:member*, Arnalds uses algorithm to control two pianos as he plays a third.[8] The computer responds near instantly to his melody, producing otherworldly harmonic progressions. There's even a 'human+machine' art festival. In 2019 the festival included Improbotics—a comedy improvisation act, whose human actors were fed lines by the GPT-2 AI as they performed, and who then worked to incorporate them into the skit.[9] Watching on, the audience performed a real-life Turing test—trying to spot the machine's contribution.

Centaur strategies

Might a similar human-machine approach permit genuinely novel, creative strategies? Human-machine teams are already arriving at the tactical level—like the 'loyal wingman' drones we saw earlier that will soon fly alongside piloted fighter aircraft. But, like the centaur chess team, the creativity unleashed by human-machine collaboration points the way towards strategically useful AIs. A machine that throws up unexpected gibberish is of limited use to strategists, but in combination with humans, new creative possibilities emerge. The machine may produce surprises. It may make startling moves that appear bolder, or more risk tolerant than a human would make. Would human strategists working with the machine go along with that? The stakes in war would be rather higher than those facing the human part of the centaur chess team. If the machine says to do something whacky, like AlphaGo did in move 37, would you go along with its judgment? You might think twice, especially now you know about its tactical brilliance and strategic naivety.

Conflicting human tendencies may ensue. On one hand, there is the tendency in some circumstances overly to trust the product of machine intelligence: psychologists refer to 'automation bias'. An example would be the experienced pilots of airlines who mistakenly rely on faulty autopilots, failing to over-ride them despite the evidence of their senses as the plane heads for disaster. Stress, time constraints, multitasking—all reduce the space available for considered thinking. In the heat of the moment, we set doubts aside and let ourselves be guided by the machine, like the captain of the USS *Vincennes* failing to interrogate the ship's targeting computer. On the other hand, in some circumstances, there is often a reluctance to trust the workings of an intelligent machine. That might especially be the case if it goes against our ingrained preferences. Stanislav Petrov famously refusing to believe his faulty nuclear warning computer is the shining example here. Which one of those two instincts predominates in any centaur strategy team is a question of psychology, not just computer science. The US Army is testing it out already—at the tactical level. In a fascinating recent experiment, the Army Research Lab set up cameras pointed at truck drivers to measure their reactions as they drove a convoy.[10] The humans could decide whether to engage their AI assistant, and then their responses to its decisions could be measured with facial recognition software and compared to their personality profiles. The sample size and the nature of the task were too small to draw any definitive conclusion, to say nothing of the ability to intuit attitudes from facial expressions. Clearly, much more work of this sort will be needed as tactical human-machine teams approach maturity.

And strategic teams? To explore how centaur strategy might work out in practice, consider again the central idea in Schelling's work—the psychological forces at work in escalation and deterrence. We might, like JFK contemplating escalation against Khrushchev, have an intuitive understanding of what happens when two humans square off against one another. But now machines are in the mix too. Will they be deterred by an enemy gambit, or will they recommend escalation? And what will the human team members make of that?

A recent RAND project set out to explore exactly this issue—by wargaming a standoff between the US and its regional allies and China.[11] RAND's researchers found that the escalatory dynamics between the two adversaries depended on the blend of humans and machines in decision-making. And, moreover, that blend need not be the same on both sides—some states are more willing to trust machines than others. As you'd expect, they also concluded that machines were less able to understand the signalling by humans, including when the latter were signalling a desire to de-escalate. The result: widespread adoption of AI systems could lead to involuntary escalation. Alarming, and again an indicator that the issue urgently needs further study.

While we await further wargames, some creative thinking is in order, keeping a well-thumbed copy of Schelling close at hand. A fascinating experiment from the world of AI cars illustrates some of the psychological forces in play when machines and humans square off. In 2018 at their research facility Jaguar Land Rover alighted on an interesting solution to a thorny problem—how to make sure pedestrians share the road with automated cars safely. Their answer was simple: put some cartoon eyes on the front.[12]

To see why, start with a pressing question: where is your AI car? They seem to be taking a long time to arrive as a commercial proposition. Today, many new cars have computer assistants for parking, lane-keeping and collision avoidance. But that's a rather limited version of driverless vehicles. Automated trucks have driven cross country, solo and in automated convoys. Driverless cars have even been used commercially by the car hire company Uber. Meanwhile, driverless cars are racking up millions of research miles on the roads and soaking up millions of dollars in investment from car and tech giants alike. Much of this driving, however, has been on broad highways in good weather. Navigating the twisting, cluttered roads of a European city in a downpour is another matter altogether. But eventually that problem will be solved—better sensors and more powerful computers should do the trick.

There's an even bigger problem, one that's really hard to code for: the unpredictability of other road users. In trials, driverless cars have found themselves being 'bullied' by human road users. These might in any case be aggressive, or risk-loving drivers, regardless of who is in control of the other vehicle. But machines are conservative drivers, by design. They don't speed, and they respect braking distance. Above all, they don't engage in a struggle of wills with other road users. That's a big problem: It's one thing not to drive like a teenage jock, but another to be intimidated by every passing idiot.

Back to Jaguar—why did they put eyes on the car? Doing so certainly doesn't help the machine understand humans any better. Rather, the goal is to fool the humans into believing that the machine thinks a little like them. As Schelling knew, that sort of mirroring was the key to coercion and deterrence. Driving was a favourite Schelling metaphor for strategic behaviour. In the game of chicken we encountered earlier, 'two teenage motorists head for each other on a highway—usually late at night, with their gangs and girlfriends looking on—to see which of the two will first swerve aside'.[13] That game, Schelling thought, was just a specific example of a more 'universal form of adversary engagement'. In fact, 'the more instructive automobile form of the game is the one people play as they crowd each on the highway, jockey their way through an intersection, or speed up to signal to a pedestrian that he'd better not cross yet'.[14] For Schelling, strategy often boiled down to coercion—making someone do something you wanted that they would not otherwise do. There were two variants—you could compel them *to do something*, or you could deter them *from doing something*. In motoring terms these were: 'Get out of my way!' or 'Don't step out!' How to do either of these? The basic idea was to threaten future punishment, and the secret of strategy, as Schelling saw it, was communication. This approach was intensely psychological, and Schelling was very interested in rationality and emotion. That's why in his colourful 'chicken' metaphor there are onlooking friends and girlfriends by the side of the road. Will the drivers' thin-skinned egos allow them to back down, overcoming their obvious fear? That didn't look like an abstract, clinical game theory to him. And what about his other metaphor, with two men chained on the cliff edge, dancing away? Again, stress and fear would be in play.

A complex dance of intentionality is at play when we drive, signalling to other drivers, perhaps with a wave, perhaps a gaze, perhaps just the movements of our car itself. Above all, there is an element of uncertainty, despite all this subtle signalling. Can I be sure that you will behave in a rational manner, or that you have judged the situation correctly? If not, I'd best err on the side of caution. The problem with AI cars is not that they are unsafe, but rather the reverse: they are *too* safe. They don't play mind games at all. Today's AI has no more ability to gauge the internal machinations of adversary leaders as a Tesla does to understand what's going on in the mind of the car in the oncoming lane. That's why Elon Musk still hasn't *quite* solved autonomous driving, no matter how often he and Tesla tease it.

Back at Jaguar Land Rover, researchers focused on the problem of the pedestrian crossing. As their car approached the crossing, a walker was preparing to step out. Should she? The huge, cartoon-like eyes on the front of the car flicked towards her. A dance of intentionality ensued—'I know that you want to step out; you know that I know...'. But the intentionality was in one brain only—the pedestrian's. Of course, the car's eyes weren't actually doing any seeing. The car had a sophisticated array of radar and cameras for that, imaging the world around it, so that the onboard AI could respond in an optimal fashion. The cartoon eyes weren't for receiving information at all, but for *broadcasting* it. The designers were taking advantage of our propensity to anthropomorphise, and to assume agency, even where that assumption isn't warranted.

Might such an illusion work for players in the RAND wargame, or even in the real-life strategy games that human-machine teams may soon play? There are certainly parallels between the Jaguar AI car and military strategy—with its dance of minds, and its element of coercion. But there is an important difference too. A military AI system isn't trying to keep everyone safe, like the driverless car. Its priority is to get its way, at an acceptable cost, whatever that might be. Unlike the car, a warbot strategist might refuse to be intimidated by humans. Perhaps if you escalate against it (like the pedestrian stepping out), it will in turn respond by escalating back. Its automated response could be devastating and happen faster than the ability of its managers to intercede. That was the point of the Soviet 'dead hand' retaliatory system—intimidation through mindless automaticity. If America attacked, a devastating counter-attack was inevitable. The AI car will stop, the strategic warbot will escalate. The fates (in the form of the coders) have decreed as much before the action begins. Once their parameters are set by their human creators, the machine won't change them in response to the enemy's posturing, in a futile bid to manipulate human psychology.

And yet, the Jaguar Land Rover experiment can still teach us something about strategy, especially to be aware of the fundamentally different approach taken by machines. The strategic equivalent of cartoon eyes would be a system that makes the enemy team *think* there is human agency involved, even if there is none, or very little. The result might be something like the stunned reaction of Lee Sedol looking across the table at the man from DeepMind, just there to move the pieces on the board, per AlphaGo's decisions. Sedol sought the human intention across the board, but walked away reflecting ruefully in defeat that he had been staring into a mirror, at his own doubts. That sort of illusion might bring some strategic advantages for the creator of the AI strategist—prompting uncertainty, caution and the desire to maintain human control themselves, even at the expense of speed.

Imagine, then, a strategic AI system—a blend of human decision-makers and machines, locked in the sort of coercive dynamic that Schelling outlined. In homage to him and Turing, let's call this system a Universal Schelling Machine, or USM.

Would adversaries treat our USM as though it was a human? Would they indulge in the sort of anthropomorphic 'I know that you know' mind-reading that the car with cartoon eyes hoped to inspire in human road users? And what about those humans working inside the USM, responding to its suggestions, or reacting to its lightning-fast decisions. How far should they trust it? The answers to all these questions must be haphazard and speculative, because there are no USMs, and not much sign of them appearing any time soon. There's rather a large gap between the sort of tactical warbots of today—Phalanx, Iron Dome, Aegis, even the AI systems of the human-piloted F-35—and the fictional strategists like WOPR in *WarGames*.

Speculative or not, there are still compelling reasons to think carefully about this question. For one, the space in-between the tactical Aegis and the USM is likely to compress as systems become more integrated and automated. That's likely to increase the level of force at which decisions are taken by the automated system, and to reduce the time available for humans to intercede. Tactical actions that happen at pace need coordinating at pace. The systems involved in i-Battle are becoming ever more powerful: faster and more heavily armed. As they do, the 'mission command' we saw earlier, where commanders delegate authority to the lowest possible level, will become supercharged. In the words of one US general, the battlefield of the near future is 'hyperactive'.[15] If that vision is realised, strategists will need a keen appreciation of how AI systems will seek to coerce, and when they will escalate. Tactical events can have strategic impact, and sometimes in unpredictable ways. Speed is collapsing the space for humans to be involved even in decisions with strategic impact.

More positively, though, where speed isn't a significant constraint and humans can have the final say, there may be scope for strategic AI to improve the quality of strategic decisions—creating something new, surprising and useful. We saw that this is possible in the tactical world of computer games and centaur chess. Moving from the toy universe to the real one though is fraught with dangers. We'll need to be absolutely sure that machines remain our faithful servants. That, however, is challenging.

The sorcerer's apprentice

It's always been tricky to present relevant knowledge to an AI. It's just not possible to capture all the salient information, including about our goals and actions. Leading AI researcher Stuart Russell refers to this as the 'King Midas' problem.[16] In Greek mythology, Midas wanted everything that he touched turned to gold. Unimaginable riches awaited, it seemed, when his wish was granted by the god Dionysus. But there was an unintended consequence—his family turned to gold, and so did all his food. How can we be sure that the strategic AI won't be like Dionysus, disastrously interpreting our wishes in an unanticipated way? Another fable makes a similar point, this time with automatons: In Johann Goethe's story, a sorcerer's apprentice is left in charge of the workshop with some chores to perform. But things go awry when the idle apprentice casts a half-learned magic spell and sets a broom to work mopping the floor. Unable to stop the broom, he splits it in two with an axe. Fans of the Disney cartoon will remember what happens next.

The problem in both fables is of unintended consequences and a lack of control. When it comes to AI, a similar problem arises, compounded by the machine's lack of common-sense understanding. How can you be sure that you represent the knowledge to the machine in a way that will ensure it faithfully sets out to achieve what you want? If you represent the knowledge to the USM in rich detail, it may be flummoxed by novelty—new and unforeseen developments arising outside its existing rules. Alternatively, if knowledge is too sparse, the machine may pursue its narrow goals with a relentless fixation that overlooks wider consequences, including those the political leadership wishes to avoid.

There is no way that *all* the salient information can be captured and presented to the machine in a meaningful way. Something will be lost in translation. Partly that's a limit of the machine, but more profoundly, the problem is human. It lies with our inability to specify goals and delineate the balance between them with sufficient accuracy. In the real world, the goals we want, the trade-offs between them, and the amount of effort we put into achieving them are in constant flux. In strategy, moreover, these are collective goals, constantly negotiated between members of society. And they change through time. There are so many variables in play, so many uncertainties. All the interlocking probabilities mean that even sophisticated probabilistic calculations will not 'solve' our battle plan definitively before the shooting starts. In terms that an economist would use, combatants' preferences are 'revealed' as the war goes on—revealed to the enemy, and also themselves.

All this gives human strategy something of a contingent, improvisational feel. That's where the scope for strategic 'genius' comes in. What if, once the fighting begins, we find we have changed our mind about our goals? What if we discover that we've left something important and unforeseen out of the reward function programmed into our Universal Schelling Machine? Will we be able to get the message through to the USM in time, before it delivers all too precisely what we asked it to? Russell proposes a neat solution. Instead of issuing the AI instructions for what to do on our behalf, we task it instead with ensuring that we are happy. That way the machine would always be checking back to ensure that it was doing the right thing, and there would be little danger of it doing unintended things. If it were the broom in Goethe's fable, it would be constantly asking the apprentice—'is that enough mopping, or do you need more?'

It's a deceptively simple solution to a thorny issue: Can you stop an AI once it's started? One popular variant on the theme is the idea of an AI that will resist being turned off, like the sinister HAL in Kubrick's *2001*. In Nick Bostrom's philosophical thought experiment, an AI is tasked with simply counting paper-clips—an innocuous task a long way distant from nuclear brinksmanship.[17] The problem occurs when the machine goes back for a recount, just to be sure it hasn't made a mistake. After all there's a vanishingly small probability it has, and since it had no instructions to the contrary, it had better make sure. Disaster strikes. The small probability of a counting error gets smaller with each recount, but never entirely disappears. Soon enough, the factory managers step in to stop the machine. Or try. Refusing to be deterred from the task in hand, the AI takes steps to stop them interfering. In a *reductio ad absurdam*, the scenario results in the destruction of humanity and the carpeting of the universe in solar cells as the monomaniacal machine garners ever more resources to assist its recount. It's just a modern reimagining of Goethe's fable.

So, if we can't accurately task the AI, and if we need to be sure of a way of stopping it, is Russell's scheme a solution? Yes—provided there's time. At the tactical level, where speed is of the essence, it's a recipe for indecision and defeat, per the precepts of Boyd's OODA loop, discussed earlier. But at the strategic level, there may more time for creative deliberation, and so more scope for continual feedback and true collaboration. Even there though, our problems remain. While Russell's solution might allow us to update the machine's reward function with our latest preferences, there's still no guarantee that we could provide them in a way the machine accurately grasps. We'd simply be updating it with the latest version of our fuzzy uncertainty. The machine would check in: 'Are you happy? What would make you happy?' Alas, we would not know.

Human-machine teams and prediction

The challenges of knowing and predicting how others will behave are formidable, even with machines and humans working as a team. But these are challenges that the American military is embracing. Since 2017, IARPA, a research agency of the US intelligence community, has been running a hybrid human-machine forecasting project, a variant of Philip Tetlock's Good Judgment Project.[18] Tetlock, recall, argued that prediction was a skill in itself, and that it was often more useful than domain-specific knowledge. He identified what he called 'superforecasters', who could apply their understanding of decision-making to almost any topic and make more accurate predictions than so-called experts in that field. Perhaps a machine could do something similar—avoiding cognitive psychological glitches, and applying its more systematic, statistics-based approach to uncertainty. If that were true, surely hybrid human-machine teams would be the best approach of all, like the hybrid teams in Dota 2, or the human-machine centaur chess players?

It seems not. According to some reports, in early matchups two hybrid human-machine teams, SAGE and Prescience, were outperformed by teams of elite human-only forecasters.[19] It's not immediately obvious why this might be—perhaps the humans in the hybrid team were the weak link rather than the machines. If so, the machines weren't able to make up the difference.

Here are some of the questions from the competition's warm up round:

> Between 6 December and 31 December 2017, will there be an attack on the drone base near Agadez, Niger?
>
> Will there be any single day in December 2017 during which fifteen or more Ukrainian military personnel are killed?
>
> Will a framework for an end to the Syrian conflict be agreed upon by 31 December 2017?

These are military salient questions. And they are binary—demanding yes/no answers, to which a probability can be ascribed. The competition featured tougher questions too, such as:

> What will be the daily closing price of gold on 27 December 2017 in USD?

To make an accurate judgment here requires a blend of cognitive skills—language processing, statistical analysis, the ability to learn from experience. This is different from the sort of prediction that might be needed down in the tactical weeds of i-Battle. The knowledge that's required there is more bounded; more like the chess board. A warbot must know whether the enemy aircraft will jink right or left. At the operational level too, the sorts of predictions are somewhat bounded: will their USM bring more forces to bear if I escalate this gunfight? But as the level of interaction becomes more elevated and less bounded, the predictions needed become more like those in the Good Judgment project and less like Dota 2. The value of human intelligence increases relative to the machine. Brute force techniques are less useful here than combinatorial and transformative creativity.

At the elevated level of strategy the questions aren't just more complicated, they're less reducible to quantitative metrics altogether. What do you value? What does honour *mean* to you? Will you carry on fighting, even after all hope has gone? At the strategic level, escalation, deterrence and compellance—the building blocks of strategy—are ineluctably human. So, in the centaur strategy unit, there's a large problem—the USM and the humans are talking different languages, which limits mutual understanding and (for the humans) trust. And moreover, it's not even clear that the computer is adding much value to the creative mix. i-War seems a very long way from Go, and even poker.

The strategic cyborgs are coming

Maybe there's another route to hybrid strategists. Traditional AI doesn't cut it, and centaur teams can't yet out outmatch the best humans. But what if man and machine truly became one? In August Cole's fun 2019 short story 'Holeshot' the US President straps into an AI system called METIS in order to tackle an impending strategic threat to the United States.[20]

> I'm wearing a haptic suit that was conceived as a mash-up of Ronald Reagan's red phone and the 'football' nuclear launch-command briefcase toted around for American presidents since the Cold War's hottest days. The old way is too slow for the threats out there. So in the blink of an eye—I'm not making this up because I can do it with a stern look—I can take the country to war.

It's *Ready Player One* meets *WarGames*. And METIS, the Multidomain Enhanced Thinking Insight System, is a true AGI possessed with remarkable empathy:

> I know METIS understands how I feel about this situation, and knows what I don't yet know, and what I'll need to decide.

When the President makes a decision the system doesn't like, METIS delivers a jolt to his body! Happily for humanity, the President, a former Marine, overrides the system and defuses the crisis. Entertaining and thought-provoking, if a long way from reality. It's one thing to combine the tactical genius of a chess computer with the strategic vision of a chess champion, and quite another to develop cyborg or centaur AI-human strategists. But a series of startling technological developments in recent years suggests of what might eventually be possible.

First, consider technologies that use AI to allow greater understanding of the biology of the human brain. The vast, intricate complexity of human brains, dense with neurons and neurotransmitters retains much of its enduring mystery. How does consciousness arise? How does that brain convert the electrical signals coming from its sensory organs into meaningful models of the external world? It's a remarkable system quite unlike any technology we've produced ourselves. Scientists have always reached for analogies between the brain and the dominant technologies of the day—clockwork, hydraulics, telegraphs, telephone interchanges, electrical circuits and latterly computers. But all do scant justice to the complexity of the real thing.

That's true too of modern AI, even when it's ostensibly based on the neural structure of the brain. Modelling the processes of the brain, and even attempting to emulate it using sophisticated mathematics and supercomputers has produced limited results. We can describe the *what* of the brain better than ever before, but the *how* and *why* remain elusive. In particular, there's a persistent gap between the microbiological processes at the cellular and subcellular level, and the psychological processes at the functional level. It's not clear that mapping the neural correlates of a particular behaviour or subjective experience with ever greater fidelity will close the gap entirely. Still, the combination of advanced brain imaging techniques and Artificial Intelligence systems has produced some startling advances that promise new understandings and hint at new combinations of human and machine.

Mind merge

Consider the scope for combining 'minds'—whether human and machine, or even human and human. Could a 'mind merge' like the sort seen in science fiction be technologically feasible? If so, that might provide a way to unite very different intelligences and bring about new approaches to strategy. Until very recently, that's been firmly in the realm of the fantastical, but no longer.

The first step is to use computers to 'read' what's going on in the biological mind. Progress here is rapid. AI has been used to analyse the vast volumes of data generated by brain scans, and to identify subtle patterns in activity connected to particular activities.[21] New knowledge about the connections of the brain, and the specialisations of particular regions has emerged. There are some dramatic illustrations of the possible. 'Locked-in' patients, physically immobile but with active conscious minds, have, when placed in a brain scanner, been able to answer yes/no questions posed by clinicians. To do so, they imagine themselves engaged in one physical activity or another, each having a distinctive brain 'signature'. Similarly, AI has been able to produce speech based on patterns of brain activity. It's even been able to 'decode' activity in the brain that suggests whether a human is lying. Mind-reading AIs like this are useful for people with physical impairments. Prosthetic arms can now be controlled purely by the power of thought, interpreted by connectionist AI and translated into robotic movement. That's achieved by reading the signals of nerves in the body, or more dramatically, by reading straight from a pattern of neural activity in the brain. In one startling demonstration, a quadriplegic woman learned to eat chocolate by manipulating a robotic arm.[22] A later demonstration was more impressive still—she flew an F-35 fighter aircraft in a simulator by the power of thought alone.[23]

The direction of travel can work the other way too—with the AI inputing information into the brain, via a sort of 'neural modem'. After all, the brain itself receives sensory input as electro-chemical signals via the nerves. These signals aren't passively received 'bottom up', but rather interpreted by cortical systems working 'top down'. The reality that we experience is a composite of new information and projections of what the brain expects. Life unfolds in a constantly revised stream of expectations, occurring in near real time. Perhaps, then, the brain can also make sense of electro-chemical signals coming from artificial agents? It can. So, for prosthetic arms, this involves connections to nerve endings that previously extended into the arm.

Now the brain relates to the arm in the way anticipated by cyberneticists—with action and reaction happening dynamically. It becomes possible to pick up an egg without cracking it through excess pressure. And, stunningly, the artificial sensations prompt real feelings.[24]

Inputing of data into the brain like this exploits the phenomenon of 'neuroplasticity'. Certain regions of the brain are functionally specialised, but brain processes typically involve a broad network of neurons, and there is sometimes scope for new regions to be recruited as part of the network. This can happen in damaged brains, for example where patients recover speech ability following damage to the regions originally associated with language recognition and production. In extreme cases, patients have even been able to compensate for the loss of an entire brain hemisphere. The left-brain hemisphere is usually home to regions specialised in language recognition and production. But patients who lose that part of their brain have nonetheless been known to recover some language ability, presumably by recruiting neurons in the right hemisphere, and forming new networks.

One striking recent example of neuroplasticity and technology working together comes from work with blind patients. Researchers used the tongue to process images from a video camera, via a tactile matrix.[25] Eventually the patient's brain learns to interpret this novel information as visual—constructing a crude image of the world that allows them to make sense of it. Visual information from the real world piggybacks another sensory system on its way to the brain. It's further evidence that the way in which we make sense of the world is not as a passive recipient of incoming information, but as an active constructor of our reality.

Now combine all these elements: the brain's ability to control machines remotely, the novel feedback coming from machines, and the mind's plasticity. New possibilities emerge. The brain controlling a third arm, for example, in coordination with the body's two biological arms, and adjusting to doing so in a way that feels natural. Would it work? People with hand and face transplants have experienced psychological difficulties in accepting the new body part as belonging to them. We don't yet have much understanding of how entirely artificial body parts will affect us psychologically—will we come to accept them as entirely naturalistic? Will we experience them as alien appendages? And what will they mean for the ways in which we conceive the world? It's one thing to imagine a new limb, like this, but what about a new mind?

The philosopher Jacob Uexküll coined the useful term *umwelt* to describe the way in which our experience of the world is largely determined by our biology. In his famous mid-century paper Thomas Nagel drew on that idea to ask 'what is it like to be a bat?'[26] Rather different, he thought, from being a human, because of the radical differences in sensory experience. Now though, neuroplasiticity allows us to contemplate modifying the human *umwelt*. The neuroscientist David Eagleman has designed a vest that incorporates tactile, 'haptic' motors which feed information to the wearer's back.[27] The sort of information is entirely flexible, and Eagleman talks enthusiastically about generating additional sensations for the wearer, depending on what's fed in. It's one thing to substitute vision for pressure on the tongue, quite another still to let the brain 'see' in the ultraviolet light spectrum that is usually invisible to us.

The potential for this sort of modification of the self remains uncertain. It's early days. But the neurological condition of synaesthesia suggests what might happen. Here, sensory information is interpreted by the brain in unusual ways, producing a sort of perceptual mash-up. The range of possible sensory overlaps is considerable. Letters on a page may be experienced as colours in the mind or taste in the mouth. Sounds can induce physical sensations in the body—you *feel* a noise, rather than hear it. Seeing someone experience something can induce the same feelings in you—a sort of supercharged empathy. Synaesthesia offers a clue as to how the brain will interact with new sensations of reality, or *umwelts*. It could be wild. And in changing our sensations, we might ultimately change what minds actually are. There may be implications for strategy—particularly when it comes to the scope for creativity and imagination.

Control of a robot arm is doable. What about control of another mind? Yes, and it has already been done. In a series of increasingly astonishing experiments, scientists have demonstrated rat-to-rat brain interfaces, human-to-rat interfaces and finally human-to-human interfaces.[28] The 'thoughts' being transmitted in each case involved motor control—with one organism directing the other to move. Step one involved decoding the neural correlates of the volition to move; step two—the most trivial—was wirelessly transmitting that data to the other organism; and came finally step three, stimulating the other organism to move. With rats, this impressive feat can be achieved invasively, via tiny electrodes that stimulate neurons in the target rat's motor cortex. With humans, while that is technically feasible it presents obvious ethical dilemmas. Deep brain stimulation is used in humans, for example in the treatment of Parkinson's disease and epilepsy, but is invasive and risky. So instead researchers used transcranial magnetic stimulation, which can produce pulses of synchronised neural activity in targeted brain regions.

How does it feel to be the body on the receiving end of a mind merge? It's still very early days, so there's much we don't know. But it might have parallels in the out of body experience, which can itself be stimulated by electrically zapping the angular gyrus, a brain region that creates an internal representation of the body. Or it might feel something like patients with 'alien hand syndrome'. This bizarre sensation sometimes emerges as the result of brain surgery that separates the two hemispheres of the brain (typically as a treatment for severe epilepsy). Sufferers retain a single, cohesive self-identity, but nonetheless are unable to explain why the motor cortex in the right hemisphere directs their left hand to do something (because the rationalising 'narrator' in their brain sits in the left hemisphere). Or perhaps it could feel like the sort of effect produced by hallucinogenic drugs like LSD, which prompt in some users a sense of their own self dissolving somehow, sometimes into a feeling of pan-consciousness with the world around.

Maybe, but the sort of 'mind merge' demonstrated so far is a very long way from a true merging of consciousness. These demonstrations amount to an undeniably impressive party trick, one that combines technologies that hitherto were established in their own right. DARPA sponsored scientists demonstrated a remote-controlled cyborg beetle as far back as 2009. You can watch it on YouTube, buzzing across a room on its wire tether, essentially being 'flown' by human operators stimulating its wings.[29] But motor control is only a small part of what the brain does. Striking though it is, a thoroughgoing mind merge will require much more. For one thing, it will involve emotions.

Emotional cyborgs

For thousands of years, western philosophy has juxtaposed emotion and 'reason' as antagonistic forces. Too much emotion was the enemy of sound decision-making, Aristotle thought, and many others have since followed suit. In modern social science the dominant model of decision-making is still the 'rational actor model'. This approach assumes that humans make decisions using some sort of systematic process of calculating odds and payoffs. Undergraduate economics courses still routinely start with the idea of humans as calculators of 'utility'—seeking to maximise the payoff from a choice by working out its 'expected utility'—the product of the likelihood of it happening and the size of the benefit if it does. That sort of maths is, most utilitarians would concur, only ever intended as an abstraction of how people actually decide in the real world. We know enough about real brains to suppose that they don't behave like a statistician armed with a supercomputer. The process by which the brain actually decides is altogether more organic. In particular it involves emotion and parallel processing. Expected utility is recognisable as the way in which Kasparov's nemesis Deep Blue actually arrives at its decisions, using computer power to make millions of probability calculations in seconds. In our search to understand the strategic implications of AI, we can learn something important by reflecting more on the distinction.

The last few decades have seen a renaissance in thinking about emotions in decision-making and behaviour. In part that's because we have better neuroscientific techniques to study their role in the brain. Perhaps in part it's because we've changed our theories of mind—the analogy of brain-as-computer has become less dominant in psychology. It's still a commonplace analogy, but by the 1980s the weaknesses of traditional AI had abundantly demonstrated to psychologists that brains were altogether more than logic machines. Whatever the reason, it's now clear that emotions are not the enemy of sound reasoning, per Aristotle, but inescapably an integral part of our reason.

Most strikingly, some unfortunate patients, including those studied by the neuroscientist Antonio Damasio, experience localised brain damage that impairs their ability to experience emotion.[30] And they were often lousy decision-makers. Why? Damasio argued, and many neuroscientists and psychologists today agree, that this was because the emotions do much of our deciding for us. If we decided like 'rational actors' in an economics model, we'd never get anything done. Life would be an endless and laborious process of cost-benefit analysis, with the added complication that the payoffs and probabilities for many choices were unclear. Instead, emotions make many of our decisions for us, or at least steer us towards certain options, perhaps even outside of our conscious deliberation. Does a situation make you feel uncomfortable or anxious? Avoid it. Feeling angry about something? Step up and confront it. Emotions act as a filter, focusing attention, and as a cognitive shortcut, framing our decisions for us.

There may be distinct emotional systems for so-called 'basic' emotions, like fear and anger; but complex emotional shades are likely to include different brain regions in subtly different combinations. Many of our emotions are responses to social situations—envy, embarrassment, *schadenfreude* and so on. Parts of the brain specialised in attributing context and meaning interact with areas of the brain that respond to reward, and parts that receive sensory information. The improvised jazz orchestra that is the human brain is constantly sensing, predicting and adjusting—with emotions as an integral component. Paul Slovic, focusing his research on how emotions shape our judgment of risk, refers to an emotional 'heuristic'. In truth, there isn't really one, but many such heuristics, or mental shortcuts. After all, the brain is a parallel processor, par excellence.[31]

A lack of emotions might explain why AI is comparatively poor at creativity and has no ability to gauge meaning in the same way we do. These are embodied and encultured aspects of our intelligence. But what happens if we create emotional Artificial Intelligence? Cyborg technologies point the way. A more profound mind merge than the motor-control version would involve projecting emotions from one brain into another. At the time of writing this book, this had not been achieved, but the constituent elements are there already.

Decoding the emotional signature of basic emotions like fear and rage is not straightforward, but it can be done. Neuroscientists have a decent understanding of neural networks associated with fear and rage. For example, multiple routes for fear processing in the brain have been found, one direct and swift, fast enough for a real emergency; another a little slower, allowing the brain time to appraise what's going on.[32] But subtler emotional hues are another matter—especially where they involve a significant social element. Mapping that sort of network state is problematic given the insensitivity of today's measuring instruments and the complexity of electro-chemical interactions involved. Transmission, by contrast, is trivial—as before, that's just moving data across the ether. What about the input? It would be fiendishly complex to replicate the full network state of the originating brain in the target. But perhaps something cruder *is* possible. In a recent experiment, scientists 'input' emotion using 'optogenetics', a relatively new technique. They were able to stimulate an aggressive response in mice, via genetically modified, light sensitive neurons in its amygdala, an evolutionary ancient brain region integral to mammalian fear and rage systems.[33] Essentially, they made the mouse furious and aggressive at the flick of a light switch. More recently still, another startling experiment optogenetically stimulated oxytocin neurons in mice via a wireless headset, this time stimulating pro-social behaviours.[34] Emotion at the flick of a light switch.

Impressive though these feats are, they are only proof of concept for the capacity to manipulate emotion. For one thing, the technique was invasive—a large contraption grafted onto the head of the mouse projected light down onto the sensitive neurons. That's certainly not the sort of thing most humans would sign up to. For another, the emotion involved, rage, is evolutionarily ancient and rather unsophisticated. It's one thing to stimulate a knee jerk, instinctive response like that, quite another to produce a more subtle emotional palette, especially one requiring contextual appraisal of the sort that involves the cortical regions of the brain. And the result is mind control, not mind merge: there's nothing here to suggest shared consciousness.

The bottom line is that shared cognition, of the sort described here, is in its very early stages. Already though, there are some intriguing possibilities for new forms of intelligence. One is extra-limbs, or control of extra-bodily vehicles. Here, the bridge from existing tech to science fiction is shortening. In *Blade Runner 2049*, the sinister corporate titan responsible for a new generation of artificial humans is blind, but like the wearers of the dot-matrix tongue, he 'sees' using a swarm of tiny airborne robots, feeding sensations straight into his visual cortex via a small modem mounted behind his ear (which, actually, is the wrong place).

It's not clear how human brains and minds will process these new sensations. Will plasticity allow the brain to make sense of ultraviolet light, or feel information about stock prices, as David Eagleman hopes of his haptic vest? Will mind merges really allow the blending of thoughts and selves? It sounds sci-fi: the Borg of *Star Trek* are famously of one mind. The supercomputers of *The Matrix* create an emotionally engaging virtual universe for the prone, incapacitated humans whose energy they are harvesting. In Netflix's *Altered Carbon*, consciousness can be transferred from one human to another. Then there's the entirely disembodied consciousness of the Puppet Master in the classic manga film *Ghost in the Shell*. But all that's a very long way from unidirectional motor control or even the transmission of aggression in mice.

Chimeras

There's one other form of 'altered carbon' that might make an impact on strategy. It too counts as Artificial Intelligence, and it too is a long way from what we currently think about when using the term. Chimeras in classical Greek mythology were fantastical creatures composed of more than one animal. Often the mix was a bizarre, fire-breathing blend of lion, goat and snake—one is described in Homer's *Iliad*. No such creature exists today, but one project comes close—in it, human-like ears were grafted on the back of mice. Google 'Vacanti mouse' to see it—the ear was actually made of cow cartilage, not human.

Genetic engineering like this raises the prospect of more far-reaching hybrid creatures, including humans. Moreover, the possible changes include those affecting the brain structure and function. Already genetic engineering has produced new pest-resistant crop strains and new farm animals with more consumable meat. That's a modern twist on longstanding farming practice of selectively breeding new types of foodstuff. But it has proved controversial where the new produce would not be feasible using traditional techniques like selective breeding, or tree grafting. And it's especially controversial where the end product is heritable—an animal or plant able to produce offspring in a self-sustained manner.

Now, modern developments in human genetics point the way to some dramatic possibilities, including the creation of wholly new 'chimeras' that would startle ancient readers of Homer. Two technologies have made a big impact here. The first is our old friend AI, running on classic von Neumann machines and doing what it's uniquely good at—recognising patterns in large datasets. The second is a technique, still fairly new, that allows scientists to cut sections of a gene and insert them into another genome, like a set of genetic scissors. The most commonly used approach here goes by the name CRISPR. Together these new techniques create the possibility of genetically engineering new forms of human intelligence. The result would be an augmented intelligence that uses computers in the design stage, but that is entirely biological. How might it work?

The first step is analysis. The human genome is a huge dataset—we have somewhere between 50,000 and 100,000 genes, of which some 25,000 code the proteins that in turn shape our cells. Each of these genes has multiple base pairs, sometimes numbering in the millions of bases per gene. The blueprint-of-you is made of some 3 billion such pairs. That's only the start of the complexity, because the contribution that each of these pairs make to you—to your body, your brain, your personality—is often far from obvious. You can see the scope for connectionist AI here, whether in cataloguing the base pairs, understanding how they in turn shape the body's proteins, or in establishing patterns between all that data and clinical diagnoses. This is exactly what geneticists are doing today—utilising the power of deep learning to understand what genes do. The dream is to use that understanding to improve our health, through individually targeted treatments tailored to our own DNA.

What sort of traits might it be useful to genetically engineer? The scope is vast. We might start with the altogether humane intention of eliminating harmful brain diseases and disorders. Some like Huntington's disease or cystic fibrosis are caused by mutations in a single gene. But that's very rare: most diseases and disorders result from many genes working in combination. The same is true of nearly all our traits. Geneticists refer to 'polygeneity'—the idea is that hundreds, perhaps even thousands of genes each contribute a little something to a given trait.[35] Physical traits like height and skin colour; cognitive traits, like intelligence, or empathic ability—all are the product of many genes working together in concert, with each making only a small contribution to the overall effect. Moreover, the same gene may be making a contribution to another trait at the same time, in combination with a different set of genes. Sampling lengths of our genome and analysing the patterns of base pairs within them can help geneticists understand what's doing what.

For over a decade, geneticists have used AI to help with the sequencing of genes and the analysis of how they relate to the character of their organism. But our understanding remains sketchy. Partly that's a technical matter: rather than surveying the whole gene, geneticists sample only sections of the code at one time. But partly it's to do with the complex way those genes manifest in cells and, ultimately, behaviours. It's not just that each base pair makes only a tiny contribution to the overall effect, but also that development—the 'nurture' part of the nature-nurture equation—also shapes the way genes are expressed as behaviours.

One important idea that's emerged from modern research is that many cognitive attributes are distributed along a spectrum, including mental 'disorders'. So we can think of a person not as schizophrenic, or psychopathic, but as lying somewhere along a distribution for that tendency.[36] Some researchers hold that a person's position along that spectrum is influenced by the mutation load in their gene—a description of how dense their code is with errors. Another important idea is that certain cognitive traits might be composites of underlying constituent tendencies. What constituent traits, for example, go into making someone more conscientious? Or a good leader? The broad idea is that if you change the code, perhaps eliminating some mutations, you change the person. But the devil is in the detail.

With this limited understanding of how our genes work, it's clearly early days for genetic engineering of humans. And light years from the sort of engineering done by the fictional backstreet geneticists of *Blade Runner*. Still, progress is rapid. We now have personalised medicines tailored clinically to the individual genes of patients. And in the last few years, geneticists have started to use CRISPR to engineer human genes. They removed patients' cells, engineered the defective bit of DNA, and then injected healthy cells back into the body. In one trial the genetic CRISPR scissors were themselves injected directly into the eyeballs of patients with an inherited form of blindness.[37] So far, so uncontentious, if not for the squeamish. But in 2018 Chinese scientists caused international controversy by using CRISPR to edit the genes in human embryos which subsequently developed into two baby girls.[38] The goal of the edit was to genetically block HIV infection. But there was a huge ethical implication: these genetic changes, unlike those in clinical trials elsewhere, were heritable. If those girls go on to have children of their own, a whole new, genetically modified line of humans comes into being. Another problem: it seems as though the edit in question didn't actually do what was intended. And even if it had, who knows what else the genes in question were also up to. For his efforts, the lead scientist was reportedly jailed for three years by the Chinese authorities.

Elsewhere in China, gene editing experiments that are also ethically contentious continue. In 2019, two pigs were born with monkey DNA present in many of their organs.[39] Earlier, their embryos had been injected with genetically modified stem cells from cynomolgus monkeys. The ostensible goal was to pioneer techniques that might be used to produce human-compatible organs for transplant to needy patients. And the chimeras did indeed have a small proportion of monkey cells in many of their organs. But there was a problem: both monkey-pig chimeras died within a fortnight. Clearly the research is very much work in progress.

Then came another, still more controversial Chinese experiment. The monkey-pig experiment could have been done with human DNA, and that's exactly what another team of scientists did, also in 2019.[40] This time, rhesus macaque embryos were injected with a modified virus, which was used as a vehicle to carry a human gene. Six of the eleven macaque babies that resulted died soon after birth; but the five survivors were a world first—partly human animals. Chimeras had arrived. And what's more, the scientists claimed some spectacular results—the chimeras' brains took longer to mature than unadulterated macaques, and they seemed to possess better short-term memory too. As an encore, the team is apparently at work with other human genes, including those associated with language development in humans.

Augmented strategists

What does this bizarre experimental work imply for strategy? A clue comes from a fascinating experiment a few years ago, when researchers discovered that chimpanzees outperform humans at a two player, zero sum strategy game.[41] The simple game required players to guess how their adversary would respond in their next move and push one of two buttons on a touchscreen accordingly. One player had to match the button of the other, while the second player had to mis-match—by pressing the opposite button. There could only be one winner (who, in the case of the chimps, would win some food, as incentive to keep playing). Then the experimenters would swap who was the matcher and who was the mis-matcher, unbeknownst to the players.

How quickly would they understand the new incentives and adjust? It turned out that chimpanzees are really good at this game. So good, in fact, they come close to the mathematically optimal response, the Nash equilibrium that we encountered earlier. Humans were not so good. The big unresolved question is why? Clearly the game contains some elements of strategy—it's adversarial, with a clear winner; and it requires prediction about what's coming. But it is basic stuff—a very narrow 'toy universe'. Perhaps it is better thought of as tactics than strategy. If so, might the chimps not be a little like the relentless pattern-recognisers of Deep Blue and less like the visionary, creative chess genius that is Garry Kasparov?

The researchers thought the difference might lie in our large, but nonetheless still limited, brain capacity—perhaps we'd lost the sort of pattern recognition skills that the chimps still clearly retained when we developed language? Possibly. There is something to the idea of a trade-off, and language acquisition may be a part of that. But an alternative possibility is that humans developed social skills that became our default way of tackling adversarial, dynamic interactions like these. Specifically, we developed that rich 'theory of mind'. Richer at least than the chimpanzees. For humans, match-mismatch, like chess and Go, is a game of intuiting what the other player is thinking. It's that familiar dance of 'I know, that you know...'. While we are busy playing this game of intentions, the chimps are calculating patterns, like the poker playing AI Libratus. Another game at which they excel reinforces the point—this time it's a simple question of short-term memory for pattern.[42] Numbers flash briefly on a touch-screen then disappear. Next, more numbers, including some of those from before. If the chimps identify what they've already seen, they win a treat. Again, their performance is better than ours.

What then if we could capture, or perhaps re-capture some of this skill? What if, rather than inserting human DNA into a macaque, the experimenters decided to try a bit of chimpanzee DNA in a human brain? Science fiction—but only just. Thus, the genetically engineered human strategist comes hazily into view. It would be another form of 'augmented intelligence', like the blend of machine AI and neuroplastic human brains. It might be designed to avoid debilitating disease, or to eliminate extreme brain disorders, like bipolarity, whose sufferers experience cycles of crushing depression and exhilarating-but-exhausting manias.

But at what cost? Again, science fiction comes to our aid. In *Gattaca*, one of the all-time great science fiction films, Ethan Hawke's character Vincent dreams of travelling to the stars. But he is excluded from the astronaut programme because he has 'normal' DNA, with all the mutations that we have now. (The movie's title spells out the letters of our genetic bases—G, T, C and A.) Sadly for Vincent, the space-going elite are all genetically perfect—by design. Undeterred, he bluffs his way into the astronaut programme by passing himself off as genetically pure, with the help of blood and body tissues from Jude Law's character, Jerome. A genetically superior Olympic medallist, Jerome is now a wheelchair bound paraplegic, thanks to a traffic accident that may have been a failed suicide bid. The film reminds us that genes are only part of the story of our lives: beware genetic determinism. When eventually Vincent outswims his genetically pure brother and flawlessly pilots the spacecraft in training, he illustrates for us that genes can't straightforwardly predict performance. What even are the genes in question here? Some might govern motor control, allowing a smooth swimming stroke, others aerobic efficiency. A complex of physical attributes go into swimming, and that's before we come to the willpower and determination to train, and to compete. Editing that into a person won't be straightforward.

If our goal were to produce better strategists via chimeras—what exactly constitutes 'better'? We know that personality is multidimensional—so which virtue should we maximise and which minimise? Life is sufficiently rich to make reducing it to the pursuit of some goals, or the maximisation of some capacity, problematic. Take a trait that seems, at first blush, unarguable: the maximisation of intelligence. That would certainly be of interest to those charged with executing strategy. There is considerable evidence that a general intelligence is measurable, via the long-established IQ test. Though not without controversy, the intelligence quotient correlates well with educational attainment, and with some metrics for life outcomes—including measures for health and employment. The test is reliable, returning similar measures through time with the same person. And there is a body of evidence suggesting a significant genetic component to intelligence too. Moreover, there is evidence that links cognitive performance across a range of seemingly disparate tasks in one generalised intelligence measure, sometimes called simply *g*. So, people who perform well at spatial reasoning will also do well at verbal and numerical reasoning, and at reaction times too—all things measured in an IQ test. Why the correlation? Again, there's an idea, still to be robustly established, that performance is in part a measure of mutation load—a measure of the frequency of mutations in base pairs on our genome.

If that's true, then one way to enhance intelligence would be to eliminate some of those mutations, perhaps via CRISPR editing-in 'healthy' DNA. Perhaps while those future geneticists were at it they could even splice in some chimpanzee DNA to boost our evolved capacity for pattern recognition. Monoglots might get a tweak to their FOXP2 gene, implicated in the acquisition of language in humans and song in birds. If we see intelligence as an unalloyed good, and mutation as one drag on it, what's the problem?

One we've already encountered is polygeneity—the idea that genes might be doing more than one thing. Weeding out one error might cause problems elsewhere. Another is that there is no ready template for perfect intelligence. For one thing, what if some of those mutations were actually making a positive contribution to intelligence? After all, that's the basis of evolution—that some random mutations confer an advantage. For another, there's that question of creativity, encountered earlier. We have no doubt that the brilliant artist, the insightful author or the genius physicist is possessed of something. But what? Why, for example, are some of the most famously creative people also unconventional? Why do some experience mental disorders like bipolarity, schizophrenia or Aspergers? Perhaps these conditions give rise to different perspectives on reality, raising the possibility of making new combinations of existing ideas, and perhaps even of truly innovative transformational creativity. Evidence that creative people suffer more from disorders like these is often anecdotal, or based on very small samples, and concentrates on the extreme end of the creative spectrum, especially on famous people—the brilliant artist, like Vincent van Gogh, often retrospectively diagnosed as bipolar; or the genius mathematician, like John Nash, Nobel prize winner, pioneer of game theory and paranoid schizophrenic. But some larger studies also suggest that sufferers are over-represented in creative employment. Some people with 'non-typical' mental processes refer to 'neurodiversity'—the idea that not everyone is wired the same way. It's that diversity which might be sacrificed in pursuit of a souped-up intelligence.

So from a narrow interest in strategy, which is our perspective here, this sort of editing might result in a less creative intelligence. The capacity to see things differently and to make imaginative leaps is prized in gameplayers and generals alike. We might instinctively feel that reducing errors and increasing *g* is the way to better strategy. But raw intelligence is not the only indicator of talent. Perhaps higher IQ contributes to the sort of intelligence exhibited by machines and chimps—pattern recognition, statistical thinking, the capacity to sequence moves that look like they will yield the best payoff. Eliminating mutations could improve the essential *efficiency* of the brain: learning, memory and the raw computational speed of information processing. But as Garry Kasparov argues, chess genius isn't just about intelligence, it's about vision and instinct. Sure, you can leverage the raw power of machines to analyse adversary gameplay, but those machines can't do strategy, no matter how fast they get. And you can improve the *g* of a human, but it will never match the computer. As Clausewitz argued when describing his genius of war, a very different sort of intelligence is what's required from commanders: one that is more instinctive, more creative and perhaps even more empathetic: one that knows its enemy and knows itself.

And so the geneticists forge ahead with their morally troubling experiments, creating bizarre chimeras, mind merges and brain-machine interfaces. The sorts of intelligence they create are certainly artificial—but perhaps it makes more sense to think of them as 'augmented intelligences' which retain the human element, including our creativity and imagination. The implications of this research won't be realised for many years, but there's already enough to suggest that today's warbots, digital and electronic weapons, are only the beginning of AI's involvement in war. Serious thinking about the military implications of augmented intelligence is in its infancy. Among the issues that will need addressing—who will be augmented, how will they interact with 'natural' humans and what happens when technology makes them obsolete.[43]

And there's a final observation that might trouble those in the west thinking about AI in war: It's no accident that all that chimera research occurred in China. If America leads in classic AI research, China is breaking new ground in biotechnology, perhaps because authorities there are more sanguine about the ethics of genetic manipulation. Recently, France declared that it would begin research on developing 'enhanced' soldiers. Acknowledging the controversial aspects of this research, the French defence minister clarified that the augmentation would be outside the body, for now anyway: 'Yes to Iron Man', as she pithily put it, but 'no to Spiderman'.[44] It's a distinction which may prove difficult to sustain. The military committee granting permission for the research acknowledged that future enhancements might be implanted in the body, but they insisted that genetic changes would be banned. For now. Which slippery terrain brings us at last to the ethics of warbots.

RULES FOR WARBOTS

Accidents will happen, but ideally not when warbots make life and death decisions. And especially not where they have the capacity to escalate violence at speeds that constrain the ability of humans to intervene. Today's warbot will be an accurate shooter—perhaps more accurate than even skilled humans. It won't experience fatigue or feel any emotions like fear and rage that might affect its performance. But we also know that it profoundly lacks our human understanding. And we know that deep learning AI is occasionally susceptible to catastrophic and unanticipated mistakes. We saw that it can be 'spoofed', so that it thinks a turtle is a rifle. Or that it blindly follows the biases in its training data, so that it returns racist or sexist results. Finally, we know that the internal workings of today's 'deep learning' AI are a black box—if you can't see what's going on as you design it, how do you know what will happen when it's unleashed on the real world? These fears animate campaigners against 'killer robots', who want to regulate or ban outright AI in battle.

'Normal accident theory' addresses the scope for mishap when systems are highly complex and interconnected—like today's AI.[1] The literature references space rockets, nuclear power plants or air traffic control systems. As the name suggests, the theory holds that accidents are unavoidable. With so many moving parts, it's inevitable that sooner or later, something will go awry. Perhaps there's been an error in design. Perhaps an unforeseen event, or a combination of events, pushes safety systems beyond their capabilities. Normal accidents often start with a small error that cascades through the complex, interconnected system. There are some landmark examples—the Challenger Space Shuttle disaster of 1986, or the Chernobyl nuclear power station meltdown, that same year. We already encountered the Aegis system shootdown of an Iranian passenger jet in 1988. Sometimes the failure is technical, sometimes human error is involved. Usually, it's a combination of both. After all, humans design and operate the system, so human involvement is inevitable somewhere along the line.

And yet we are surrounded by complex systems, many of which make life and death decisions. Air travel might seem hair raising: there are many thousands of huge planes in the air simultaneously. In peak times, one lands at Heathrow airport every few minutes. But air travel is remarkably safe. Occasionally, of course, a plane does crash, and the vivid destruction and loss of life ensure it makes a large impression in people's memories. And yet statistically there is no safer way to travel. Normal accidents? Not really; air travel is highly reliable.

Can we reduce the scope for a 'normal accident' with warbots?[2] And if not, how much control should we surrender to them? What sort of redress is there when something does go wrong? Why can't we just ban them, if there's any danger at all? Much of the discussion about warbots concerns the ethics of using them and the need for regulation. But too often this discussion floats free of the sort of analysis that's brought us this far—about the deep connection between AI research and military utility, and the impact of autonomous weapons on both tactics and strategy. Unless the discussion of AI regulation is grounded in the realities that are driving their development, there's little prospect of anything emerging from the interminable discussion but well-intentioned utopianism.

Human ethics in war

Perhaps the most obvious solution to the warbot challenge is to provide our warbots with an ethical framework to follow in combat. That's what we do for our human soldiers, after all. Maybe the simple solution is to craft, and then upload ethical instructions for robots. Rules, in other words. That's what Asimov was driving at with his rules for robots that we encountered in the introduction, and that's where we'll end this book, with my suggestions for rules for warbots, tailored to the unique demands of coercive violence.

If only we could code human ethics and then upload them to the machine, to ensure it does what we want. That, however, won't be easy. Even without autonomous machines, there are tremendous challenges in shaping a collective vision of how 'we' ought to behave, including in war. And then, once the shooting starts, there are additional difficulties in ensuring that those we have charged with enacting violence actually behave that way. All those difficulties exist with warbots too. And there'll be additional factors to consider—such as how to enact justice when behaviour deviates from what we anticipated.

But let's try. What goals should we set for our machine? Like many ethicists, I think there's a deeper origin to our morality than is captured in prevailing group norms, or even the exigencies of the moment. We can better understand the challenges to 'coding ethics' for warbots by exploring those deeper roots. Specifically, looking at the deep evolutionary origins of morality, helps to explain the common ethical features that emerge between otherwise very different human cultures.[3] At the same time, those evolutionary origins allow us to understand the scope for cultural variation in ethics, and also the tensions that emerge within any particular group's moral code. Once we've done that, we'll have firmer foundations for our putative rules for warbots.

There's a sense in which ethics constitute our groups. We are how we behave. The influential anthropologist Clifford Geertz suggested that society is rather like a spider's web of mutually agreed conventions.[4] Of course, other things may constitute our group too—that we share a contiguous piece of territory, or that we are related by kinship, or by ethnicity, for example. But groups are also shaped by narrative—especially by the stories we tell ourselves about who we are. That story is often moral; it's about how we should behave.[5] Many groups have some sort of foundational myth that sets out appropriate behaviours. If the hold of group identity is sufficiently powerful, these stories even shape the boundaries of the imagination—what we think of as possible. A whole battery of psychological processes work to orient us within this collective identity. Our group really matters to us, perhaps because in our far distant history to be cast out from the group was effectively a death sentence.

Part of the group story involves goals—what we want, and what we should do to get it. The group wages war, and our culture tells us why and how. In political science, scholars have emphasised the material dimension of war—that we fight for territory, for resources, perhaps for security. All true. But we are also motivated by meaning—a less tangible goal, perhaps, but often no less potent. Even a 'realist' like Henry Kissinger, one often associated with the narrow, objective and materialist way of thinking about international affairs, acknowledges the point. The national interest, he once argued, was about identity. That's why the tired old debate in international relations scholarship rumbles inconclusively ever onwards—should you pursue an 'ethical foreign policy', or be guided by other, perhaps more cynical precepts? Separating the two strands is often impossible, because interest, identity and ethics are deeply entwined. It is more realistic to acknowledge, as did the arch realist Kissinger, that there's a moral dimension to identity. That's one reason that the Greek historian Thucydides still resonates today, centuries later, with his attribution of war to 'fear, honour and interest'. What is honour, if not a shared story about how we ought to behave?

Justice and war

So far, so straightforward. But what *is* honourable? Alas for warbot coders, there's no consistent set of rules that captures how one should behave in war. There are, however, some common ethical themes that can be abstracted from the rich detail of everyone's particular historical morality fable. One of the most famous is the idea that we should 'do unto others as we would have them do unto us'—sometimes referred to by ethicists as the 'golden rule'. Many cultures across human history have had some notion of this golden rule. Today we see this in norms about fair treatment for prisoners, or in the prohibitions on atrocity and torture.

Evolutionary theory offers a framework for understanding why the golden rule is so common. It has its basis in the idea of reciprocity, which evolutionary psychologists suggest is the very basis for human society itself. For human groups to extend in size beyond small bands of inter-related families, we needed some way of trusting others. Robert Trivers suggested that reciprocal altruism provides that social glue—if I can trust you to do me a favour in turn, I might be prepared to help you out now, even if that comes at a personal cost, or risk to me.[6] Of course, trust isn't a given—people have to earn a good reputation, hence it pays to be generous. After all, 'what goes around, comes around' as the English idiomatically say, which is just another way of expressing the golden rule. There's a related idea in the writings of twentieth century ethicist John Rawls.[7] He thought we should treat others as though we were living behind a veil of ignorance. We make our moral choices, knowing what the consequences would be in all respects except one: we are ignorant about which person we ourselves are once the choices are made. When the veil is lifted, we find out—so we have a self-interest in being as fair as we can to all. And there's another echo of the golden rule in the classical liberalism of John Stuart Mill, whose liberal vision allowed people to do whatever they wanted, without censure from others, except insofar as their behaviour made other people worse off.[8]

What all these visions have in common is the idea of some rule for how to behave. There may be no one to enforce the rule, but this sense of being obliged or duty-bound underpins what's referred to as a 'deontological' approach to ethics—one that entails a duty. And even without anyone to explicitly enforce the rules, we sometimes internalise them enough that there's no need for an external monitor, or even, ofttimes for a sense of duty. If generation after generation grows up in an environment where social reputation is important, those who internalise that and automatically behave cooperatively are likely to thrive. It pays to have a good reputation. It feels good to be generous.

Related to this reciprocal altruism is the idea of reciprocal punishment—that we should, for example, exact 'an eye for an eye, a tooth for a tooth', per the Hebrew Testament. This is a very different idea of reciprocity—of tit-for-tat, or responding in kind. There's a powerful logic here too: if we get a reputation for being a reliable punisher, potential adversaries might be more effectively deterred. We might even get a favourable reputation within the group, especially if we carry out the punishment on behalf of others, and especially if we do so at cost to ourselves.

This deontological, or duty-bound, way of thinking about ethics has found its way into the loose tradition of morality that governs the use of violence in war. War necessarily involves violence—that's part of its inescapable nature. How can it be morally permissible to inflict suffering and death on other humans? Over time, the answers to these questions have been distilled into what's now known as the 'just war tradition'. The tenets of the tradition are encoded in modern international jurisprudence—in international humanitarian law that governs the conduct of war, in the UN Charter that sets out obligations for member states, in various international legal treaties regulating the weapons of war, and in international law prohibiting genocide, torture and other crimes against humanity. It's far from straightforward to enforce these laws—there is no international police force like those existing within states; and while there are *ad hoc* war tribunals, and even an International Criminal Court, not every state has signed up. There are bodies to monitor compliance with various treaties, as the International Atomic Energy Agency does for nuclear matters; but there's usually no obligation to be a member, and no automatic enforcement when in breach of a treaty's provisions. What remains is rather hit and miss—the intervention of states on a case-by-case basis, as and when they deem it in their interests. And, as a further sanction, the soft prohibition of ostracism by the international community—of distancing members from the circle of right-thinking states.[9]

Actually, this is rather analogous to the situation facing hunter-gatherer communities in human pre-history. How did they ensure that everyone behaved in the absence of judges and police? Clearly, a lack of formal institutions did not mean that there was no justice at all. Rather, it meant that groups of humans would have to cooperate together to negotiate what was right, and to enforce the collective view. Bands of men cooperated to violently impose the group norms. Richard Wrangham, whose research spans primatology and anthropology, suggests that this proto-judicial role was sufficiently influential that, over time, co-operators proliferated in a population, and selfish, sociopathic males were eliminated.[10] Here is the military 'band of brothers' in embryo: a small group that works together, sometimes at great personal risk, for mutual benefit. Practical deontological ethicists, in other words. And now, on a macro level, similar forces shape the behaviour of large groups in international politics. Who can you trust to uphold international ethics? The repetition of behaviour gives some indication. If states have proved reliable allies before, perhaps they will again. If they've behaved justly in earlier wars, perhaps that too is a guide for subsequent behaviour. A shared commitment to common values is another indicator. Institutions may be formal, as with the International Criminal Court, or the Convention on Conventional Weapons, or they may be informal, and created in the moment, like the 'coalition of the willing' assembled by the US to invade Iraq in 2003.

Ad bellum and in bello

What's important, per the just war tradition, is that groups should feel they have a 'just cause' for going to war, and also that they should behave ethically within it. Theorists of just war distinguish between justice in the causes of war (*ad bellum*, in Latin) and in conduct within it (*in bello*).[11] It's a rather neat distinction that many human cultures might not recognise as being so clear-cut: whether to fight someone and how to behave towards them would often be part of the same equation. Still, over time, modern states have settled as common practice that they should distinguish between lawful combatants, who can be killed, and those who cannot, like civilians, injured soldiers and prisoners. The precise constitution of those groups has changed over time and across cultures, but the broad idea of distinguishing in this manner has been commonplace in many cultures.

As for the just cause, the tradition is quite restrictive—the only legitimate reason to fight is if you are attacked. A modest extension allows groups to fight to protect others. Two further conditions are often mentioned—that groups should only embark on war as a last resort, having exhausted all other peaceful options; and that there should be some reasonable chance of success. The problem with all these sensible and morally commendable rules is that they are largely subjective. How do you know it's the last resort? How can you gauge your prospects of success accurately? Clearly your enemy disagrees with your judgment. Is it self-defence to attack first, if you fear an imminent attack from a rival? If so, what would constitute 'imminent'?

You might think these moral precepts are less relevant to our discussion of warbots—insofar as that discussion focuses on the tactical aspects of AI, like duelling swarms of robots. But they *are* salient if we broaden the discussion to include strategy, where AIs help shape the framework for human decisions. Or if we consider the adversarial cyber-sparring of rival AIs—below the threshold of war, but with the potential to escalate rapidly. And they are certainly salient if we look ahead to the future of augmented intelligence and the merging of AI and biotechnologies. So perhaps we shouldn't be too sanguine about focusing purely on the ethics of AI *in bello*, as does most of the literature about the ethics of 'killer robots'. There's a particular risk where speed is of the essence—as with missile defence systems. Our warbots may have the ability to initiate action if they sense an attack, and then to escalate once it's underway.

The just war tradition also governs the conduct of belligerents once the fighting begins. The main deontological responsibility here is to distinguish between those involved in fighting, who are legitimate targets for your violence, and others who are not. This is something of a modern convention. Many hunter-gatherer societies have the concept of a warrior, but it's often expansively defined, so that most males are expected to participate—and there are sometimes elaborate rituals to mark the rite-of-passage as a boy becomes a warrior. These foster norms about cooperation and shared risk.

Nowadays though, the army is a profession for soldiers—an increasingly small subset of the population in modern states. And there's an expectation that violence will be limited to these uniformed personnel. States are required to discriminate between those soldiers and non-combatants, which in the chaos of battle is not always easy. And, moreover, they're required to deal humanely with the enemy—providing medical assistance to casualties and ceasing hostilities with those who surrender and treating them humanely.

That's the theory, but there are many practical problems. Sometimes it's easy to tell who is a soldier, if they are wearing a uniform and fighting for a government. But what about in guerrilla warfare, where the enemy can look just like a regular civilian, as is sometimes done deliberately—using civilians as cover. Or what about those who make a contribution to the fighting power of the enemy, but don't actually do the shooting? Is it acceptable to target a factory that makes bombs? What about the bank that finances the factory? To help answer those questions, military ethicists sometimes adopt a very different ethical yardstick than deontology.

The ends justify the means

There's an alternative moral framework that might be used to govern human behaviour in warfare—that of utilitarianism, or consequentialism. Here, the morality of an action is weighed by its outcome. The essential idea is to maximise the greatest good. An important question here is—the greatest good *for whom*? Another, more practical, question is how you might measure goodness in the first place. This has never really been answered definitively, though suggestions include physical wellbeing, happiness, wealth or life satisfaction. Which one of these matters most depends partly on the context. In warfare, we might think in terms of life and death. Accordingly we would weigh our action depending on the numbers of people that will die, and choose the best course accordingly.

But whose life counts? Now your consequentialist logic rubs up against the deontology we explored previously. In war we strive to protect our own, even if that means killing the enemy. We might also strive to discriminate between enemy soldiers and civilians, with soldiers' lives counting less in our macabre calculus. But the maths gets trickier if we think about weighing our own group's lives against the lives of enemy civilians. Imagine, for example, that the enemy is hiding somewhere in a densely populated town. If we don't kill the enemy, they will surely attack us, perhaps killing our civilians as they do. Can we flatten the town in a massive bombardment that will kill the enemy, and also many thousands of innocents besides? It would save lives on our side, but is it justified? Some would view this sort of attack as disproportionate, and in so doing they would be making a consequentialist judgment.

But tweak the arithmetic in this scenario, and there comes a point where you might think the attack justified. As we reduce number of civilians in the town who will die, and increase the number of our own people who will be defended by killing the enemy, there comes a tipping point where the morality of the operation shifts. This is the doctrine of 'double effect', which holds that it's morally permissible to do something that you can foresee will have adverse consequences, provided the benefits from doing so are sufficiently large. The ends justify the means. If the consequences are sufficiently dire, we can sometimes suspend our deontological duty of care to every single innocent life.

And so, we mount an attack on a target, knowing beforehand that innocent lives will be lost.

But where is that elusive tipping point? It's a largely subjective judgment, and one that has varied from culture to culture, and war to war. These are real problems. In recent wars fought by western liberal states against non-state adversaries, the enemy has blended with the civilian population, hiding among them. Guerrillas travel in cars with children, or hold meetings in hospitals and schools, making it harder to target them. Western states have developed a technologically sophisticated way of war, using weapons that are (when functioning correctly) highly precise and discriminate. All the same, difficult choices have to be made, using this awkward blend of deontological and consequentialist logics.

The movie *Eye in the Sky* revolves around this sort of ethical dilemma—a drone strike on a terrorist hideout is about to happen, when into view comes a winsome young girl, selling bread. Should the attack go ahead regardless that she will be killed? Adding to the drama, it looks from surveillance footage as if the terrorists are about to launch an imminent attack. Should you kill the small child? In the movie, the decision is referred ever upwards to the highest reaches of government. Luckily there is time for that to happen. In real life, by contrast, these sorts of decisions are gamed out beforehand and captured in rules of engagement for the military commanders making life and death decisions. Sometimes lawyers are involved, but the go/no-go decision is delegated to the tactical level.

Now try another scenario. Imagine that you are commanding an assault on an enemy position. You know that the unit mounting a frontal charge at the enemy guns will incur very heavy casualties—most of the assault force will die. But in doing so, they will absorb the attention of the enemy gunners while another team sneaks round the flanks and takes the position.

Once that's done, your forces can break through enemy lines. In weighing that sort of decision, you're again making a consequentialist judgment—what's the greater good here? You are performing the same sort of consequentialist calculation as before, but this time weighing the lives of your own soldiers against the good that may come from the breakthrough. Do you not have a duty of care to each of them, all of whom have a right to life? And is this setup not predicated on some very shaky predictions of what might happen in the future? How can you be certain that the breakthrough will bring a large payoff?

Scenarios like these explore the tensions in just war theory. Warfare inevitably demands consequentialist judgment. Strict deontology is a recipe for pacifism. But there are profound limits in our ability to predict. How can you be sure that your attack will kill X number of civilians, or calculate the scale of damage that will ensue if you don't target the terrorist? A stark example is the employment of nuclear weapons against Imperial Japan in 1945. The decision was made ostensibly to bring about Japan's unconditional surrender, and thereby avoid a potentially bloody invasion of the Japanese islands by land forces. But how many lives would the attack have cost? How hard would the Japanese have fought? The basic problem with all these judgments is that military power is revealed *in the moment*, as it is employed.

Clearly then ethical matters get more complicated when deontology and utilitarian logics are blended. And that's not the end of just war either. Ethicists also concern themselves with justice *after* conflict—*jus post bellum*. Especially salient here for our discussion of warbots are notions of restorative justice and reconciliation. These focus attention on questions of accountability, punishment and rehabilitation. All of those are human attributes. Breathless media coverage of AI sometimes gives the misleading impression of human agency—as, memorably, when Saudi Arabia granted citizenship to a robot called Sophia.[12] But we really are far from creating sentient machines who might have a plausible claim to legal personhood and whom we might hold to account after a conflict. The only way to get satisfactory justice for the performance of machines is to hold humans to account.

There is a final precept from just war theory that might be applicable to warbots—if they are considered *malum in se*, or evil in themselves. That is, weapons that are so far beyond the bounds of civilised norms as to be unacceptable in any conceivable circumstance. Why? There's a certain amount of redundancy there: evil weapons violate one or other of the principles we've already covered. They are disproportionate or indiscriminate, or they cause unnecessary suffering. Some ethicists have argued that nuclear weapons constitute such a weapon—if they were ever used, they could not be anything other than immoral, given the sheer scale of suffering that would ensue. Perhaps a case could be made that warbots are evil in themselves because they abrogate responsibility for killing to something that is inhuman. It's the very lack of personhood that makes them evil. To automate killing in this manner is to strip human dignity from victims.

The concepts of *jus post bellum* and *malum in se* both return us to the inseparable human essence of morality in war. It's not just that there are separate, irreconcilable moral logics in war, but that these originate in the human condition.

Just war and machines

What about machines? They lack our biology—with its resulting motivations, for food, security and sexual reproduction. Machines did not evolve—and nor did they evolve our moral framework either—our need for justice, our desire to behave honourably. They lack our social organisation, where fear of violence strengthens social cooperation and produces expectations of how one ought to behave. On one hand, all that might make them more reliable combatants—they won't escalate if they feel slighted or go on a rampage motivated by rage and revenge. They won't freeze in fear or flee the battlefield, leaving comrades in the lurch. They won't experience post-traumatic stress, or suffer high suicide rates, addiction to drugs or alcohol. That is, they won't experience moral harm, and they might not inflict it either—certainly not for their own reasons. A warbot is likely to be more accurate, proportionate and discriminate than human soldiers, at least if you consider the matter tactically, as a question of directing firepower.

Nonetheless, their lack of understanding creates scope for great moral harm. Tactical warbots won't understand the context they sense—is the man picking up that weapon to shoot, or surrender? Nor will AI systems escalate or be deterred for the reasons humans do. That might be good, given, for example, that humans sometimes escalate from anger that they later regret. But it is patently not so good if a machine is undeterred by the sobering fear of things getting out of hand, like that experienced by Kennedy and Khrushchev over Cuba.

Ultimately, the machine is amoral; neither good nor evil. That's what's meant by lacking personhood. Warbots are not motivated by anything other than the instructions we give them. Still, they are more than a simple mirror of our motivations, whether morally sound or otherwise. There's a distorting effect produced by outsourcing our actions to AI. We can't code and upload our morals, because, as we've seen, they are philosophically inconsistent and always in flux. We attempt to freeze convention and tradition in international law, but it's an imperfect process, not a final settlement. Anyway, in practice, as the fighting unfolds, we sometimes change our minds about what's appropriate. It's not that laws are silent in times of war, as Cicero suggested, but that they change.

The biology of morality matters, and machines just don't have it. We know from neuroscience that different brain systems are more active when we decide on the basis of deontology than for consequentialism.[13] The former activates more emotionally sensitive areas, the latter more abstract reasoning. We also know there are personality differences that shape moral judgments—we're all individuals, and the moral particulars we follow are negotiated collectively. So it matters who codes the morals, and when they do it. In particular, the way in which an ethical dilemma is 'framed' or presented to humans matters. Describe a situation one way, and humans tend to reason deontologically; change the language, and they follow a different logic. And then there's the 'frame problem' for machines, or more broadly the challenge of translating our wishes into machine code. The most we can hope is that our warbots will capture the thrust what we intended at the time we issued our instructions. That may or may not be enough to ensure they do what we want.

When they don't, there's no scope for punishment, and no restorative justice. The warbot doesn't care if we turn it off, no matter what science fiction says. That lack of direct justice is a problem in other fields—for the AI car that crashes, or the medical AI that misdiagnoses. As with those fields, we can still find humans to blame and hold to account. If there's a faulty part or biased training data perhaps we can trace that to the company that made the warbot and put in some quality control measure. Perhaps there is corporate, rather than individual responsibility, lying with the bureaucracy that commissioned the weapon, tested it and exercised the capability. Ultimately, perhaps a share of responsibility lies with the society that collectively permitted lethal autonomy to be developed and deployed.

There's nothing new here, on one view. Today, many militaries field complex systems that can go wrong, as anticipated by normal accident theory. That already includes AI technologies—since parts of the 'kill chain' are already autonomous. The pilot of an F-35 fighter is only one element in a far larger system whose computers help identify and select targets. The human member of the human-machine team surely has diminished responsibility if the AI part makes an error, and safeguards fail. That's what happened with the USS *Vincennes* shootdown of Iran Air flight 655. There have always been situations where the individual who acts has diminished responsibility. They are doing the best they can with their understanding of what is going on. Responsibility for what goes wrong lies with the system as much as the individual.

Perhaps then, what's new with AI warbots is not the broad principle of corporate or collective responsibility, which we've always had. The Defense Department can accept collective responsibility for a faulty warbot, as they might with other faulty systems that cause death. But can we accept the loss of final control that comes from abrogating the ultimate decision to an amoral agent—the final trigger pull, or the decision to escalate? A human may be partially sighted and might not share our precise moral framework. But they look like us, they have our basic biological drives, and we can empathise with them. They preserve our humanity, even in warfare. Not so the machine.

Regulating warbots

If that makes us uncomfortable, it's up to us to regulate warbots. Other weapons are regulated, including by international treaties, so why not AI? Alas, I think that will prove almost impossible, and that instead we'll fall back on unilateral 'rules' for our warbots that will maximise their utility while remaining consistent with our societal values.

One, possibly insurmountable, challenge is to define exactly what's being regulated. We saw that warbots have a history reaching back to the acoustic torpedo and proximity fuse. Nowadays, AI systems are deeply woven into the fabric of modern armed forces. They automate processes throughout the armed forces—not just the weapons that do the actual shooting. Harnessed to the latest connectionist AI, there's much more to come in the years ahead. New technologies are emerging right now: hypersonic cruise missiles and loitering mini-warheads; nuclear powered torpedoes, able to prowl the deep almost indefinitely; solid state laser guns, capable of downing pilotless drones as they approach ships at sea. AI is integral to all these exotica. Do you regulate by weapon type, or by the underlying AI technology?

Doing either will be tricky. There's great variety in the physical platforms, but also huge variation in the underlying code and in the capabilities it generates. AI isn't a single technology, like a nuclear bomb. The basic science of a thermonuclear bomb is functionally identical no matter who makes it, which constrains the variation from one model to another. Not so for AI code. There are a bewildering variety of approaches to Artificial Intelligence, and even considerable variation in what the term means to different people. For some it conjures visions of distant science fiction capabilities, for others, it's an automated thermostat, or the satellite navigation in your family car. If you can't rigorously define AI, how can you hope to regulate it?

Even if you could come up with some agreement there, there's still the large question of 'dual use' applications for much AI. A nuclear bomb isn't much good for anything else. An ICBM might be useful for getting satellites into space, but there's no need to launch that off a submarine, or to develop MIRVed warheads to sit atop it. Stealth technology may have spurred research into exotic material compounds. But most stealth comes from the shape of the aircraft, and there's not much call for that in civilian transport. Weapons technologies often have a narrow military utility. By contrast the same algorithm that can tirelessly scour X-rays searching for tumours can tirelessly scan drone footage searching for targets. AI is adept in searching through mountains of data, or making logical inferences, carefully weighing probabilities. That's all very general, with a huge range of applications, both military and civilian.

There is more scope for effective regulation when it comes to 'augmented intelligence' via the combination of AI and biotechnology. That's because deep public unease about genetically altered humans extends further than their military potential. Still, there's unease about genetically modified crops too, and that's not been enough to stop them. It's hard to argue against changes that eliminate brutal inherited diseases; but where does one draw a line if that sort of engineering is allowed? Even less likely is agreement on regulating electronic bio-hacking, as with 'mind merges' using cortical interfaces. Today, DARPA can solicit proposals for a neo-cortical interface without violating any international convention, and that's likely to remain the case.[14]

Alternatively, you might also try to regulate AI by its effect, rather than by its constituent parts—rather like banning nuclear warheads above a particular kiloton limit, or forbidding particular categories of weapons that are deemed indiscriminate, like anti-personnel mines and cluster munitions, both now regulated by widely accepted treaties. But the effects of AI are just as varied as the code and hardware that constitute it. AI enhances 'fighting power' not just via the control of fighting platforms, but by the generation and analysis of information, and by decision-making and assistance across all levels of war, from a small gunfight to a confrontation between nuclear powers. Regulators will struggle to define what effects should be banned.

Perhaps you could limit the ban to one part of the AI's role? That's what some in the campaign against 'killer robots' want to see, whether because of the lack of human dignity that comes from a fully-automated weapon, or because an automated trigger puller doesn't understand context. Both logics are flawed. Even human shooters have a partial understanding of the situation, which is often shaped by AI systems, working in the intelligence and targeting process. Anyway, enforcing that sort of ban would be nearly impossible where platforms are only a small software update from fully automatic, and there's little way of onlookers knowing otherwise. It's a ban so fragile it would scarcely be worth the trouble negotiating. States might even sign up to it, rather like they did to the well-meaning Kellogg Briand Pact of 1928 whose signatories committed not to use force to resolve conflicts of any nature.

Cheats prosper

One reason for scepticism is the ease of 'defection'—cheating on any international restrictions. Even if you could somehow agree to regulate or ban military-salient AI, overcoming the problem of definition, you'd need to monitor compliance. And unlike nuclear weapons research, there's little tell-tale signature from an AI facility to suggest nefarious activities inside. With a nuclear programme, there are indicators to watch out for at every stage of manufacturing. First, aspirant nuclear powers need the raw materials—notably un-enriched uranium. Then they need large, dedicated plant to enrich the uranium so that it's suitable for a bomb, for example by using centrifuges. They also need scarce expertise, and perhaps blueprints of designs from elsewhere. And later, if they are successful in building a prototype, they'll need to test it. Underground test explosions can be detected with seismic monitoring, and atmospheric tests create fall-out. Lastly, they'll need a delivery system. Ballistic missiles might not themselves be prohibited, but producing them is a warning sign that nuclear weapons might not be far behind. In short, nuclear weapons programmes are hard to hide, as several aspirant states have discovered.

AI is rather different. Certainly, there are bottlenecks in expertise, and a reluctance on the part of many specialists to work with defence—but that's not insurmountable, especially in economies where experts can be directed to work with the military. An AI research programme also needs huge computer and electrical power, because training vast artificial neural networks of inter-linked GPUs sucks up a lot of electricity. But computer power becomes ever cheaper, year-on-year, and electricity can be had much more readily than enriched uranium. As for augmented intelligence, the Chinese experiments with various chimeras shows the difficulty in monitoring: this is research that can be conducted in biotech labs that do less controversial genetic research, like non-heritable gene editing, or genetic modification of foodstuffs.

Perhaps nuclear weapons are the wrong analogy. The signature of a biological warfare programme is rather smaller, and we have successfully crafted regulations to outlaw biological weapons. Except it's not clear that we have. The nerve agent used in the 2018 attempted assassination of Sergei and Yulia Skripal, and the radioactive isotope polonium-210 used to murder Alexander Litvinenko in 2006 suggest that some states still have an interest in violating the conventions, where they can get away with it. In the days of the Cold War, the USSR maintained an extensive biological and chemical weapons programme—creating new toxins and new ways of weaponizing them as aerosols. Stockpiles of thousands of munitions, and large research institutes evaded detection by the US for decades, until the collapse of the Soviet Union. There were certainly suspicions, and even some evidence from defectors—but proving anything concretely was another matter. When the Coronavirus struck the US in 2020, President Trump wasn't alone in hinting darkly that a military research facility in Wuhan, the Chinese epicentre of the outbreak, might have been responsible somehow. A former head of the British intelligence services echoed the suggestion.[15] The assessment of the intelligence agencies was otherwise and, moreover, hinting darkly about Chinese biological weapons may have served the President's domestic political agenda. But regardless of the accuracy of the claim, the episode highlighted once again just how difficult it is to monitor compliance with arms agreements. How would you know what China was doing in its biological research facilities without intrusive verification? And even then, could you be sure that you'd visited all its sites?

And when it comes to AI weapons there will be a powerful incentive to cheat. It's relatively easy, for one. Acquiring the state of the art is more feasible than for nuclear weapons or weaponised toxins and viruses. Unlike the mass state-controlled effort to develop nuclear weapons, as with the wartime Manhattan Project which built the bombs used against Japan, most AI research happens in corporate or university settings. Even better, the players are keen to publish their findings as quickly as possible—the competition to demonstrate the latest breakthrough is intense. Developing militarily useful AI still involves extensive state level funding, but as DARPA has demonstrated over decades, that doesn't need to be via direct ownership of the scientists or equipment involved. Nowadays the main impediment to cheating is the ability to repurpose civilian AI for military purposes. Hard, but certainly not insurmountable and very tricky to spot.

So states have the means to 'defect' from any regulatory regime. And they certainly have the motive. Warbots are poised to have such a dramatic impact on fighting power that states can hardly be sanguine about the prospect of a rival getting ahead. Scholars refer to a 'security dilemma' where the fear of missing out on important military developments prompts states to rush into acquiring them. This in turn alarms their rivals, who embark on their own armament programme for fear of being left behind. An 'arms race' ensues—just like the one for battleships between Germany and the UK ahead of the Great War, or between the USSR and US in ballistic missiles during the 1950s. The security dilemma has a long pedigree. Thucydides, the Athenian general and author of the first modern history, described how the Peloponnesian war was touched off by the fear that Athens' rising power precipitated in the rival city-state of Sparta. Nowadays concerned analysts in Washington fashionably invoke the 'Thucydides trap' to describe the prospect of rising Chinese power. Part of that alarm is prompted by China's undoubted gains in Artificial Intelligence.[16] Some are concerned that the Chinese are closing the gap on American research, undoing the technological advantage that has underpinned American military dominance over its rivals for well over half a century. It's fair to say an arms race is already underway, fuelled by the security dilemma.[17] So much for the desire on the part of concerned scientists and humanitarians to ban 'killer robots'.

Offensive instability

The competition is already acute, in part because there is so much uncertainty over what AI weapons can do. One aspect of that is the offence-defence balance. As we saw, tactical AI might shift that balance in favour of the attacker. If so, the appetite for cutting-edge warbots will be especially intense. Offence dominant weapon systems promote instability. If going on the attack confers a big advantage, everyone must keep their finger near the trigger, like rival gunslingers at the OK Corral. That drives arms races, and further impedes the prospects for regulation.

Conversely, defence dominant weapons can exert a stabilising influence on international affairs. If you know your attack will likely be beaten back, you might be less inclined to launch one. Submarine launched nuclear missiles are the ultimate defensive weapon—extremely hard to locate, and able to deliver a devastating counter-attack. Rational leaders must carefully consider the prospect of retaliation before doing anything too provocative against nuclear powers. Certainly, that's been the view of states who've acquired nuclear weapons to deter aggressive local rivals. With a retaliatory nuclear bomb, you don't need to match the conventional military power of an adversary—just make sure there's enough doubt in their mind about your response to any attack. Thus nuclear weapons promise defence on the (relative) cheap.

But offence-dominant AI weapons might be less stable. They might even undercut that 'second strike' nuclear capability. That's happened several times in the nuclear era, with weapons that threatened to destroy the enemy's ability to retaliate. In the 1980s, for example, America's medium range Pershing II missiles greatly alarmed the Soviet Union because they could reach Moscow in minutes from launch sites in western Europe, and so effectively decapitate the Soviet leadership, leaving it unable to order retaliation.

Could warbots undercut defensive measures to the extent that even a nuclear deterrent is threatened? Possessing a sufficiently large, or well concealed nuclear arsenal provides considerable insurance. It's well-nigh impossible to launch a surprise attack and obliterate the many thousands of nuclear weapons held at readiness by the US or Russia. But the scores available to China, still less the handful of missiles deployed by the UK and France aboard a single patrolling submarine are another matter. It is at least conceivable that these might be found and targeted by new technologies. These states have placed very few nuclear eggs in very few baskets. As for the more threadbare deterrent fielded by North Korea, or nascent programmes elsewhere—these could be altogether more vulnerable to AI attack.

Improbable, perhaps. But at bare minimum, the prospect of a system like this is alarming for states that have grown used to relying on a secure deterrent, and the stability it brings. Additionally, destroying missiles is only ever part of the equation—what if your AI swarm could decapitate the enemy leadership, perhaps like the surprise attack in the *Slaughterbots* film, produced by anti-warbot campaigners? There, tiny micro drones assassinate individuals that they identify using facial recognition—technologies that are increasingly mature.

I might be wrong about the offence-defence balance. Maybe nuclear submarines will continue to evade detection. Maybe spoofing will throw the attacking swarm of robot assassins off the scent before they can find their leadership targets. Online too, perhaps the initiative enjoyed by AI attackers will not prove decisive, as AI cyber-defences like DARPA's Mayhem anticipate weaknesses and patch themselves before rivals can pounce. Perhaps too encryption will continue to win the race against codebreakers, protecting defender's communications from paralysing attack. Still, states can't afford to be sanguine about the prospect of offence dominant AI. And if so, they must be particularly sensitive to the relative power of rivals. Adding to the uncertainty is awareness that a marginal qualitative advantage in AI may prove decisive in combat. An AI will defeat a human pilot if it cycles its decision-making a hundred times faster. But in turn it will be destroyed by a rival AI that thinks 101 times more speedily.

In this uncertain, offence-dominant environment, states will never have enough AI to feel secure, either quantitatively, or qualitatively. In the 1960s, US Defense Secretary Robert McNamara famously articulated a new nuclear doctrine that he termed 'assured destruction'. An M for 'mutually' was added soon after by commentators—MAD. There was no need, McNamara argued, endlessly to pour national wealth into ruinously expensive nuclear systems. All that was required were enough missiles to ensure that only a suicidal madman would deliberately attack you. In practice, that still meant a lot of missiles, but at least McNamara capped some of the rampant spending. Today, this is the sort of minimal deterrence that underpins British strategy. Why keep spending on huge stockpiles of weapons, locked in an endless competitive cycle with rivals? Alas, in the world of AI weapons, that might not be enough.

In the early years of mutual deterrence, one nuclear theorist described the standoff between nuclear superpowers as 'the delicate balance of terror'.[18] There were certainly some close shaves and a healthy dollop of luck. Actually, though, the balance of terror has proved pretty robust—precisely because of the defensive quality possessed by nuclear weaponry. Fear played a considerable part in keeping the peace, but so too did the technology itself, with its ultimate defensive backstop of assured retaliation. Thousands of missiles and warheads were stockpiled on each side. Yet it didn't matter how many weapon systems each side developed, or in what quantity: there were never enough to be confident of a disarming first strike. This time round, however, the balance of terror might be much more delicate. Automated systems are certainly scary because of their potential to go wrong—a 'normal accident' leading to inadvertent and unproportionate escalation, or indiscriminate killing. But they're also scary by design—because they dramatically increase fighting power, especially of attacking forces.

Means and motive

So states have both motive and means to acquire AI systems and to resist efforts to regulate, still less ban them. So far, many advanced defence producers have made reassuring noises about keeping humans involved in decision-making. The British government, for example, has been clear about never developing full AI weapons. The small print though is revealing—for the UK, an 'autonomous system is capable of understanding higher-level intent and direction'.[19] As we've seen, that's well beyond the state of the art for today's AI. It's more like the human-like Artificial General Intelligence of Hollywood movies. Anything less can be considered merely as an 'automated weapon'. It's an artful get-out that allows the development and deployment of highly capable AI weapon systems that will soon force humans well away from the loop.

And reassuring noises aside, advanced states haven't banned them. It's hard to see them ever doing so. In the United States, the Congressional National Security Commission on AI neatly summarised the challenge. AI, the Commissioners noted, was the quintessential 'dual use' technology; it would transform all aspects of military activity; large scale proliferation was to be expected; and defending against AI capable adversaries without employing AI yourself would be an 'invitation to disaster'. Their conclusion: the US should lead a coalition of democracies to develop military-salient AI.[20] With a touch of hyperbole, we saw Russia's Vladimir Putin declare that the country that masters military AI will control the world. If so, he ought to be concerned, because Russia lags far behind the US and other leading AI research centres, like the UK and Israel. And it lacks much of the societal infrastructure that has proven so conducive to research in all those places. Still, hype or not, Putin certainly captured the febrile atmosphere of the day—including the anxiety about lagging behind. Such uncertainty is driving much of the race to develop warbots and associated capabilities. If you don't know what the state of the art is for your rival it's impossible to say, 'we've got enough now'.

With so much uncertainty about AI systems and warbots, anxiety rather than overconfidence is the order of the day. Also in the mix are very different conceptions of how societies should develop and use AI. Some states are less concerned with protecting individual rights, whether at home, or possibly in battle too. It's all a recipe for investment and experimentation, for arms races and instability. And not, sadly, for restraint, transparency and regulation.

Rules for warbots

If international regulation won't work, perhaps self-regulation is the only way ahead. And if governments won't step up, maybe AI scientists could 'self-regulate' to stop the creation of warbots. Certainly, there's deep unease among many researchers about contributing to the development of warbots. This transnational 'epistemic community' of like-minded scientists might mobilise itself—influencing what companies and research institutions that fund them choose to work on. We already saw Google employees protesting their employer's decision to support the Pentagon's work on Project Maven. More recently, we saw companies including IBM, Amazon and Microsoft step away from facial recognition AI, in response to the Black Lives Matter protests. This isn't inconsequential: norms matter and can act powerfully to shape cultures of national security. They are a key driver of how violence gets used—that's why states see assassination by poison as odious, even though it's feasible and highly discriminate. Even where there is no legal prohibition, norms have real world consequences. Where they do, laws can follow—enshrining liberal attitudes in international humanitarian law, as with the conventions banning the use of anti-personnel land mines and cluster munitions.

But the prospects are slim. Researchers don't really form a cohesive epistemic community of like-minded concerned citizens. And both corporate interests and the security dilemma are powerful counterweights. Amazon and Microsoft both bid to supply highly classified cloud-based computer services to the intelligence community; Palantir stepped in to work on Maven. University researchers participate regularly in DARPA funded projects. And elsewhere, computer scientists are hard at work in the research labs of traditional defence contractors, like Boeing, Lockheed and BAe Systems. There are, once again, echoes here of the nuclear era, and the Pugwash committee of concerned scientists campaigning for nuclear disarmament. It didn't work for them, and it's far from clear that the situation has changed now. Reluctant scientists and the inherent conservatism of armed forces might slow the development of warbots for a while—but that's hardly reassuring.

What then can we hope for? I think the only prospect for effective AI that works consistently with our values is to craft our own rules, not wait for international agreement. It matters greatly what sort of military AI we want as a society. We won't be able to escape the logic of the security dilemma; but nonetheless we might be able to square our moral concerns about machines that kill with the imperatives of national security.

It's time to revisit Asimov's rules to see whether they might help frame rules for warbots—not as laws, or even a framework agreed by all states, but perhaps something we might unilaterally use to guide the development of our own warbots. As a reminder, here they are:

- A robot may not injure a human being or, through inaction, allow a human being to come to harm.
- A robot must obey orders given it by human beings except where such orders would conflict with the First Law.
- A robot must protect its own existence as long as such protection does not conflict with the First or Second Law.

Those are fine rules with which to start a philosophical debate, or a discussion of sci-fi classics. Perhaps they might even be applied to AI on the factory floor or steering a driverless car. But they're useless for warbots, whose goal is to coerce, using force if necessary. Earlier, I suggested these alternatives:

- A warbot should only kill those I want it to, and it should do so as humanely as possible.
- A warbot should understand my intentions, and work creatively to achieve them.
- A warbot should protect the humans on my side, sacrificing itself to do so—but not at the expense of the mission.

These rules foreground the need for accuracy and humanity in killing. Take Rule One: the warbot should only kill those I want. We could instead have specified that the warbot should only kill those it needs to, in order to achieve the goal we've set it—leaving the particulars to autonomous 'mission command'. But that is rather fuzzy and open ended. More tightly coupling the goals and the actions reduces the machine's autonomy to decide for itself what it will take to win. But that's a price worth paying to ensure our moral values are adhered to.

One danger is that we might not know who we need to kill to achieve our goals, and so set either too permissive or too restrictive criteria for the warbot. If we get the calculation wrong, the machine may not visit enough destruction on enemy forces to achieve our goal. It would have to come back to us for further instructions, rather than escalating on its own. This is a restriction of its autonomy, and returns us to the security dilemma—what if our less squeamish adversary doesn't put that mechanism in place? What if we can't get through to the warbots to update their parameters in real time?

Might there not be a tendency to overshoot, giving more permissive instructions to compensate for any doubt about what the mission will take? Possibly, but that at least will be *our* tendency, not the machines'. And perhaps that's better than giving the warbot our goal, but not setting the upper bounds of violence, and then seeing it unleash carnage far beyond what we anticipated. Plus, it's less of a restriction on autonomy than either the famed 'human in the loop' or 'human on the loop'.

And there's another reason not to fear hamstringing ourselves with Rule One. In many crises, there may be time to adjust. Yes, speed is critical at the tactical level, and perhaps even as combats are knitted together into a sequence, as operational art. But at some stage, different qualities come into play—creativity, ingenuity and foresight become possible. The OODA loop is a tactical proposition, designed by Boyd with dogfighting pilots in mind. You can certainly apply it to higher levels of war, but only as a looser analogy for adversarial decision-making. Scale and distance create scope for strategy, and for thinking carefully about how much violence to use, and against whom. It's not yet clear precisely where the firebreak between a tactical fight and thermonuclear war is, but it's there somewhere.

So, under Rule One, good instructions for the warbot would include some desired end state, both in terms of what we expect to have happened (a mission accomplished, an enemy surrendering) and in terms of who we expect it to attack. The clause about 'humanely' is intended to prohibit gratuitous violence—it reinforces the idea of using the minimal force necessary, and of avoiding cruelty or atrocity. Machines can't really commit atrocity by themselves—they lack our very human motivations of anger, revenge, or the personality types that include sadism and psychopathy. These spur atrocity in the heat of battle. But it is possible that they act with disproportionate violence. Requiring them to act humanely is simply to emphasise the need to use as little force as possible against those we've identified as targets.

The first clause of Rule Two is a straightforward lift of Stuart Russell's idea about machines having to satisfy us, in order to avoid the 'Sorcerer's Apprentice problem' that afflicted Mickey Mouse. The machine repeatedly checks back to see if it has correctly understood our intentions. Again, there's nothing new there—human subordinates do that too, under mission command. The command of forces in war is a two-way street, or should be, if things are working well. Time and restricted communications obviously impede effective communication, but again there's that spectrum between tactical activities, where speed may be of the essence, and more operational ones, where there is scope for interaction, and where—anyway—there is a need for more than raw speed to bring satisfactory results.

What about the second clause advocating creativity? It is a reminder of the importance of ingenuity in 'mission command', whether using human or machine subordinates. We are outsourcing the fighting to agents, with the hope that their decisions are qualitatively better than ours, or the enemy's. Again, with tactical systems, there's less scope for this creativity: the action is more tightly determined by physical characteristics of the warbots and the battlefield situation. Raw computer power will be decisive. If there's any creativity here, it's what Boden called 'exploratory'—rapidly searching through the possible moves ahead. But the moment there is space for reflection, we need warbots that can add more—perhaps one day even the transformational creativity that is the hallmark of human genius.

Why would you want that sort creativity in a warbot? After all, too much imagination might mean the warbot stops faithfully following our intent and does its own thing. Not in the sense of Hollywood AI, perhaps, where machines are essentially ersatz humans, with their own motivations and imaginations. But still, even a relatively mindless machine might get a little too creative for comfort. How much improvisation does it take before a machine becomes essentially insubordinate? This is a problem for any sort of mission command. The extra twist with machines is their very different cognitive structures make it harder for us to anticipate, as we might do for human subordinates, by empathically imagining ourselves in their shoes. The search for safe and explainable AI is an attempt to address this challenge. Given the opacity of neural networks, that research sub-field has its work cut out. The other difference between human and machine subordinates is in responsibility—you can certainly blame a machine, but it won't mind in the slightest, and nor will its behaviour be shaped by the anticipation of future censure.

Still, despite the challenges, we want creative warbots precisely because of the gap between instructions issued aforehand and reality. The whole point of intelligence in machines is so that they can adapt deliberately and usefully to developments. That's essentially the definition of intelligence, and creativity is the cherry on top. It's this flexibility that distinguishes intelligence from automaticity. So there's a tension in Rule Two. Finding the right balance between creativity and efficiency, including checking in with us, will not be straightforward. A tense international standoff is perhaps not the ideal place to deploy a warbot with a curious streak.

And then the third rule. My loitering munition has no desire for self-preservation. That's one of the great attractions of uninhabited platforms. Humans can sometimes do likewise, of course—but it takes considerable motivation to produce that level of altruistic and sacrificial behaviour. We can code machines to take actions that will do everything possible to safeguard human combatants—a logical extension of the 'loyal wingman' concept. But should we? As with all combat, there's that underlying tension between the consequences of action and the duty of care to each individual. In the wars of recent decades, there was considerable debate in the UK about the application of EU human rights legislation to the battlefield. EU law enshrines the right to life—no one, the law states, shall be intentionally deprived of their life. Clearly that's not feasible in battle.

That's true of enemy combatants—the goal in combat is to compel the enemy to back down, killing them if necessary. And it's also true of our own forces: commanders have to balance risk to life against the objective of a mission. This is what makes the military profession unique—new recruits understand the concept of command and the attendant risk to their life. They may be ordered into battle in dire circumstances with small odds of survival. No responsible commander sets out to risk life cavalierly, and all commanders have a responsibility to do the best for their subordinates, including in providing the best possible equipment and training. But ultimately the only way to avoid all risk is to stay in barracks. And so the duty of care demanded of commanders does not extend to the right to life for their subordinates. So it is with the machine—preservation of life will be one of its goals, but like human agents, there will be some trade-off: under what circumstances should it preserve life? And whose life?

In Afghanistan, as the war against the Taliban dragged on unremittingly, American commanders introduced the notion of 'courageous restraint'.[21] The idea was that US troops would accept greater risk—to the success of a combat operation, and perhaps even to themselves—in a bid to reduce the risk to Afghan civilians. In doing so, they might advance the cause of the overall mission—winning over the local population. Using force discriminately and using the minimal amount of force required have long been staples of counterinsurgency doctrine, but applying them in practice, in the heat of combat has always been challenging. That's especially so where your side enjoys an overwhelming firepower advantage, and where your own society is strongly averse to casualties.

Did it work? Did the decision to use less force expose American soldiers to greater risk? If so, was that worth it to achieve the overall mission—winning over the population? It's difficult to know. The number of civilians killed by aircraft dropped significantly. And yet, there was also a dramatic increase in the number of civilians dying in combat. That's because the tempo of combat increased. Some analysts speculated that aerial attack by the US and its allies had suppressed Taliban military activity. Moreover, the Taliban themselves were responsible for killing many civilians in combat. Free from the fear of air attack, they stepped up their operations, and more civilians died. From a consequentialist perspective, courageous restraint arguably increased the risk to the very non-combatants it was intended to protect. Unintended consequences: but consequences nonetheless.

As ever when it comes to weighing consequentialist ethics, establishing simple causal relationships is difficult if not impossible. And what's true for humans is also true of warbots—after all, they are simply our agents. It's going to be just as tricky to find the right balance between fighting hard to achieve the mission and exercising the duty of care owed to our own personnel, still less our ethical responsibility to others on the battlefield. One relative advantage for warbots is that they might be more accurate than human combatants. Another is that in exercising restraint, our warbots don't need to be courageous. We will surely be more willing to expend them to protect lives, just as we will be more willing to expend them to achieve the mission. Disposability may even be a feature, especially of the robot swarm.

So Rule Three is a reminder that combat is risky, and that soldiers accept those risks. But 'courageous restraint' in Afghanistan serves as a warning that translating abstract principles like these into action is not straightforward. Finding the appropriate balance between risk and reward involves subjective judgment. Rather than rigid guidelines for warbots, the rules suggest that our arrangements should be pragmatic, and flexible—adjusting to the needs of the moment, in a way that is broadly consistent with our societal values.

My rules suggest a way ahead for our relationship with warbots. It may be impossible to ban them outright, or even to regulate them by underlying technology or combat performance. But actually, there's no real need for new international regulation to govern their behaviour. The compromises inherent in just war are already encoded in international humanitarian law today—adapting them to the era of warbots need not be an insurmountable problem. As with human combatants, warfare involves careful judgment—about who is a target, and how much force you can use in pursuit of your goals. There are real problems in answering those questions. The doctrine of double effect is particularly problematic—fudging consequential and deontological logics. But the challenges ahead have more to do with us, and less with the machines we build. There is no simple way to upload ethics, because to some extent our ethics are contingent. Yes there are cross-cultural commonalities, but ethical particulars are negotiated in the moment, by societies with particular goals, attitudes to risks and to technologies.

In the final moment, the warbot will be deciding how to behave on its own—autonomously. After all, that's always been the attraction of AI weapons. But autonomy is always constrained, whether for human or machine. Human agency is shaped by our genetic inheritance and development. It's also shaped by social factors—the group we are in, and the authority figures we follow. So *who* is responsible when you do something bad? Criminal law essentially shrugs at these philosophical conundrums. The courts assume that you are an agent, responsible for your own behaviour, albeit that sometimes there is evidence of diminished responsibility—a plea of insanity, for example. And, moreover, the law assumes you are the same person through time—the seventy-year-old languishing in jail decades after his initial crime is fundamentally the same agent—responsible for those earlier actions and liable for the punishment. If not, you'd be punishing an innocent man. In battle too, the human is an agent—responsible for their actions. There is command and military discipline, lawfully constituted, and accepted by all soldiers on their recruitment. Their autonomy is constrained insofar as they can be ordered to do things they would rather not, including risking their own life. But 'following orders', as we know famously from the Nuremburg Trials of Nazi criminals after the Second World War, is not a legitimate excuse for immoral behaviour in wartime. The soldier is an agent, but one whose autonomy is constrained, sometimes severely.

The warbot is not an agent in the same sense. It has no legal personhood—it is not a moral actor, required to exercise its ethical judgment.[22] When it makes its calculations on the appropriate balance between objective and casualties, it's not using its own judgment—but ours. We are the responsible agents. And if it captures a version of our ethical reasoning at a given time, we can hardly say later to the lawyers, 'look, we've changed our minds. We wish we'd said something different.' If we can hold the long serving jailbird accountable for his dubious moral reasoning many years after the crime, we too can be held accountable for our choices. And for our oversights too—we know about the frame problem, and normal accident theory. We know the machine won't understand precisely what we intend, regardless of how sophisticated our AI gets. If, knowing those things, we still go ahead with warbots, following the grim and inescapable logic of the security dilemma, we can hardly hold ourselves blameless if things go wrong. We chose to do it. The warbot doesn't want anything, and it isn't troubled by our moral dilemmas.

The best we can do is craft rules that reflect those tensions and uncertainties. Isaac Asimov's rules were designed to stop a sentient, human-like robot running amok. My rules, by contrast, are designed for AI that is fantastically clever in some respects, but astonishingly stupid in others. Ultimately, they're designed to hold humans responsible for the machines they build. We are entering an era of Artificially Intelligent warfare, fuelled by incredibly powerful computers. But we made these machines and we are responsible for them. As the empathic President Kennedy once argued, 'man is still the most extraordinary computer of them all, whose judgment nerve and ability to learn from experience make him unique'.[23] Let's hope so.

POSTSCRIPT TO THE SECOND EDITION

It's been a year since the initial publication of *I, Warbot*, longer since I submitted the manuscript. AI is a fast-moving field and so too, it turns out, is strategic affairs. Two huge events have dominated global agendas: the ongoing struggle against Covid-19 and the Russian invasion of Ukraine. I'll come back to those to see how they might impact my argument.

In AI research, there have been important developments too. Just recently, Open AI teased the release of its new image generator, DALL.E 2. When provided with a text instruction, this AI system returns an image, often incredibly striking and creative. Want an otter painted in the style of Vermeer's *Girl with a Pearl Earring*? No problem! All rather surreal, but what does it suggest about the state of the art in AI research, and what are the implications for conflict and strategy? If you were a professional illustrator, perhaps you'd shift uncomfortably in your seat. To me, this is another striking example of what Margaret Boden termed 'combinatorial creativity'. That is, the machine produces new blends of existing styles that make compositional sense. The output can be novel and surprising, but when you see them collectively, there's a sameness to these images, like inferior knockoffs of an artist you already know. I think the most striking creativity on show here is the imagination of humans issuing the prompts—

and of course, it's human minds that find artistic meaning and pleasure in the machine's output. Perhaps the takeaway from DALL.E 2 is that much real novelty and value arises from the blend of human and machine intelligence. I can readily see military applications, both tactical and strategic, for this sort of human–machine team, along the lines I described in this book.

There have been some other big advances in this past year. DeepMind caused a stir with its algorithm that can predict protein shapes emerging from DNA strands—one of the great challenges in biology and another stellar illustration of what is possible within the existing paradigm of deep learning: huge data harnessed to tremendous computer power leads to the optimisation of a tricky problem. So too when the French start-up NukkAI unveiled an expert-level bridge-playing AI. Winning at bridge requires many of the same attributes as poker—skill and an ability to reckon with luck. It's yet another challenging 'toy universe' with echoes of strategy, now conquered by AI. Meanwhile, back in London, DeepMind created an AI that writes computer code at a competitive level. In programming tournaments for humans, its AlphaCode finished in the middle of the pack, hinting at a future where AI systems will write and execute their own programmes, perhaps even designing their own AIs. And their own weapon systems too? This again approaches the stuff of science fiction: there's a great Ted Chiang short story about a time when superintelligent agents do science that far surpasses the understanding of humans. Perhaps self-coding AI biotechnologists of the future might seem similarly otherworldly to their human creators.

One final, stunning, development emerged this year from Google's AI research labs. Their 'Socratic' models can reason deductively on the basis of sensory information—visual, auditory and natural language—updating their assumptions in a way that looks eerily human. So, an algorithm describes in English prose what it senses, and then another reasons using those descriptions: I've seen a car, and I've heard an engine—perhaps now I'll see a driver. As ever, this is different in some important respects from human reasoning, with our distinctively embodied, emotional and social cognition. But it's a reality check for anyone who thinks that connectionist AI has plateaued with video games or autonomous driving.

All these recent developments haven't quite prompted me to change my mind about AI as a strategist. I still subscribe to President Kennedy's line with which I closed this book—that humans are 'the most extraordinary computer of them all'. And yet. Progress in AI continues to be rapid. Today, we have AI that can create exciting art and thought-provoking prose. And we have AI that can integrate sensory information and then reason about what is happening. Surely the strategic implications are only just beginning to unfold.

In some ways the conflict in Ukraine looks rather traditional. There are large armies, equipped with industrial quantities of armour and artillery. There is brutal urban combat. There's not much evidence of the cyber *blitzkrieg* that many anticipated, although that doesn't necessarily mean that cyber warriors on either side have been idle. And we are clearly a long way from the sort of swarming, mosaic warfare between vast fleets of autonomous warbots that I outlined. Above all, it turns out that the vaunted Russian military, a decade into modernisation and riding high from earlier battlefield successes, is both old-fashioned and inept. The appalling morale of its soldiers and its outmoded Soviet-era kit have factored in combat rather more than its adoption of high technology.

But for all its echoes of earlier, industrial warfare, the conflict in Ukraine says important things about the coming era of autonomy. For one thing, the battlefield is saturated with sensors, as never before. Real time imagery gave the Ukrainian defenders early warning of Russian moves and allowed precise targeting. Some of that was done by one of the standout weapons in this war, the basic but highly capable Turkish-made Bayraktar drone, destroyer of scores of Russian heavy tanks and armoured vehicles. Ukraine's infantry also deployed an abundance of small, commercial drones to conduct reconnaissance and targeting—and then used the combat footage as effective propaganda online. And on the ground, another war-changing weapon, infantry portable anti-tank munitions, proved highly effective against Russian armour, itself poorly utilised. Some of these were old weapons drawn from NATO stockpiles, but there were newer ones too, including the American-supplied Switchblade loitering munition. Like the munitions that featured in the fictionalised combat of the near future discussed earlier, Switchblade is a human-in-the-loop system, but only by design, not necessity.

There were hints of more AI technology behind the scenes, in the presence of NATO surveillance aircraft looping ceaselessly along the EU-side of the Ukrainian border. Machine learning might not have been dogfighting in the skies over Kyiv, but it was certainly parsing mountains of electronic information hoovered up by all those sensors and then connecting it to the decision-makers. Meanwhile, facial recognition tools identified soldiers, living and dead. Online too, AI was there in the relentless social media battle of bot versus bot. There were even some unconvincing deepfake videos, including one of Ukraine's president improbably conceding defeat. So, this devastating war, still unfolding as I write, combines elements of human conflict that we know already—including appalling atrocities by Russian soldiers—with a glimpse of what's coming. Not much imagination is needed to foresee a battlefield where distributed and partly autonomous systems allow persistent surveillance and the rapid direction of firepower. It's here already.

I wrote *I, Warbot* while the Covid-19 pandemic put swathes of global society into lockdown, killing millions and damaging many more. The origins of the virus remain mysterious and there's great controversy over the possibility of an accidental lab leak, more so given the additional twist that a naturally occurring virus could conceivably have been engineered to be more potent. Whether or not this particular virus *was* engineered like that, it certainly could have been. The rapid development of novel vaccines owes similarly to advanced biotechnology. The pandemic then, like the war in Ukraine, serves as a bridge to a rather disturbing future—one where the 'artificial' part of artificial intelligence need not necessarily be a traditional computer. I find that as troubling now as when writing about chimeras in the book. Together, both recent events highlight the urgent need for serious reflection about the connection between science and warfare. If, as I believe, international regulation of AI will be challenging, we need at minimum to ensure that our own technologies serve the interests, including the moral interests, of our liberal societies.

NOTES

INTRODUCTION: A CALL TO ARMS

1. *Tass Russian News Agency*. 'Putin stresses whoever takes the lead in artificial intelligence will rule world', 1 September 2017, https://tass.com/society/963209.
2. McFarland, Matt. 'Elon Musk: "With Artificial Intelligence we are summoning the demon"', *Washington Post*, 24 October 2014.
3. Cellan-Jones, Rory. 'Stephen Hawking warns artificial intelligence could end mankind', *BBC News*, 2 December 2014, https://www.bbc.co.uk/news/technology-30290540.
4. Musk, Elon et al. 'An open letter to the United Nations Convention on Certain Conventional Weapons', 21 August 2017, https://futureoflife.org/autonomous-weapons-open-letter-2017/.
5. Asimov, Isaac. *I, Robot*. London: Harper Voyager, 2013.
6. Hoffman, David E, *Dead Hand: The Untold Story of the Cold War Arms Race and Its Dangerous Legacy*. New York: Doubleday, 2009.
7. See especially, Banks, Iain. *The Player of Games*. London: Orbit, 2005.
8. Minsky, Marvin. 'A celebration of Isaac Asimov', *The New York Times*, 12 April 1992, https://www.nytimes.com/1992/04/12/business/technology-a-celebration-of-isaac-asimov.html?pagewanted=all&src=pm.
9. Clausewitz, Carl Von. *On War*. London: David Campbell, 1993.

HOW TO BUILD A WARBOT

1. On which see Payne, Kenneth. *Strategy, Evolution, and War: From Apes to Artificial Intelligence*. Washington, DC: Georgetown University Press, 2018.

2. Lieberman, Matthew D. *Social: Why Our Brains Are Wired to Connect*. Oxford: Oxford University Press, 2013.
3. Damasio, Antonio R. *The Feeling of What Happens: Body and Emotion in the Making of Consciousness*. London: W. Heinemann, 2000.
4. Frankl, Viktor E. *Man's Search for Meaning: The Classic Tribute to Hope from the Holocaust*. London: Rider, 2004.
5. Adams, Douglas. *The Hitchhiker's Guide to the Galaxy*. London: William Heinemann, 2014.
6. See especially Keller, Laurent, and Elisabeth Gordon. *The Lives of Ants*. Oxford: Oxford University Press, 2010.
7. Modern researchers too have been inspired by insect intelligence. See for example, Beer, Randall D. 'The dynamics of adaptive behavior: A research program.' *Robotics and Autonomous Systems* 20, no. 2–4 (1997): 257–289; Brooks, Rodney A. 'Intelligence without representation.' *Artificial Intelligence* 47, no. 1–3 (1991): 139–159.
8. Overy, Richard J. *The Bombing War: Europe 1939–1945*. London: Allen Lane, 2014.
9. Wiener, Norbert. *Cybernetics: Or, Control and Communication in the Animal and the Machine*. Second ed. Mansfield Centre, CT: Martino Publishing, 2013 and for an overview Rid, Thomas. *Rise of the Machines: The Lost History of Cybernetics*. New York: WW Norton, 2016.
10. Ashby, Ross, and Paul R. Stein. 'Design for a Brain', *PhT* 7, no. 4 (1954): 24.
11. Tetlock, Philip E. *Expert Political Judgment: How Good Is It? How Can We Know?* Princeton, N.J.: Princeton University Press, 2005 and Tetlock, Philip E., and Dan Gardner. *Superforecasting: The Art and Science of Prediction*. London: Random House, 2015.
12. The best biography is Hodges, Andrew. *Alan Turing: The Enigma*. London: Burnett, 1983 and for his solution see, Turing, Alan Mathison. 'On computable numbers, with an application to the Entscheidungsproblem', *J. of Math* 58, no. 345–363 (1936): 5.
13. On Shannon, see Gertner, Jon. *The Idea Factory: Bell Labs and the Great Age of American Innovation*. New York: Penguin Press, 2012.
14. On von Neumann, see Poundstone, William. *Prisoner's Dilemma: John von Neumann, Game Theory, and the Puzzle of the Bomb*. New York:

Anchor Books, 1992. On both men, see also Gleick, James. *The Information: A History, a Theory, a Flood*. London: Fourth Estate, 2011.

15. Robinson, Heath. *Heath Robinson's Second World War*. Oxford: The Bodleian Library, 2016.
16. The original statement is online at 'A proposal for the Dartmouth summer research project on Artificial Intelligence', 31 August 1955, http://www-formal.stanford.edu/jmc/history/dartmouth/dartmouth.html.
17. Wittgenstein, Ludwig, translated by D.F. Pears and B.F. McGuinness, *Tractatus Logico-philosophicus*. London: Routledge, 2003.
18. Simon, Herbert A. *The Sciences of the Artificial*. 3rd ed. London: MIT Press, 1996.
19. Chomsky, Noam. *Aspects of the Theory of Syntax*. Cambridge, Mass: MIT Press, 1965.
20. For a summary from one of the design team, see Nilsson, Nils J. *The Quest for Artificial Intelligence: A History of Ideas and Achievements*. Cambridge: Cambridge University Press, 2010, chapter 12, 'Mobile Robots', pp. 162–180. Nilsson's book is far the best account of projects and personalities at work in early AI.
21. See Salisbury, Emma. 'A cautionary tale on ambitious feats of AI: The Strategic Computing program', *War on the Rocks*, 22 May 2020, https://warontherocks.com/2020/05/cautionary-tale-on-ambitious-feats-of-ai-the-strategic-computing-program/; and Roland, Alex, and Philip Shiman. *Strategic Computing: DARPA and the Quest for Machine Intelligence, 1983–1993*. Cambridge, Mass: MIT Press, 2002.
22. Lenat, Douglas B. 'CYC: A large-scale investment in knowledge infrastructure', *Communications of the ACM* 38, no. 11 (1995): 33–38.
23. Turing, Alan Mathison. 'Computing machinery and intelligence', *Mind*, Volume LIX, Issue 236, October 1950, pp. 433–460, https://doi.org/10.1093/mind/LIX.236.433.
24. Baard, Mark. 'AI founder blasts modern research', *WIRED*, 13 May 2003, https://www.wired.com/2003/05/ai-founder-blasts-modern-research/.
25. Nilsson, *Quest for Artificial Intelligence*, p. 323.

DEEP WARBOTS

1. On which history, see O'Mara, Margaret, *The Code: Silicon Valley and the Remaking of America*. New York: Penguin Press, 2019.
2. Davies, Joshua. 'Say hello to Stanley,' *WIRED*, 1 June 2006, https://www.wired.com/2006/01/stanley/. For details see, Thrun, Sebastian, Mike Montemerlo, Hendrik Dahlkamp, David Stavens, Andrei Aron, James Diebel, Philip Fong et al. 'Stanley: The robot that won the DARPA Grand Challenge.' *Journal of Field Robotics* 23, no. 9 (2006): 661–692.
3. Kim, H. Jin, Michael I. Jordan, Shankar Sastry, and Andrew Y. Ng. 'Autonomous helicopter flight via reinforcement learning', In *Advances in Neural Information Processing Systems*, pp. 799–806. 2004.
4. For the history, see Cobb, Matthew. *The Idea of the Brain: A History*. London: Profile Books, 2020.
5. McCulloch, Warren S., and Walter Pitts. 'A logical calculus of the ideas immanent in nervous activity', *The Bulletin of Mathematical Biophysics* 5, no. 4 (1943): 115–133.
6. Selfridge, Oliver G. 'Pandemonium: a paradigm for learning', *Mechanism of Thought Processes: Proceedings of a Symposium Held at the National Physical Laboratory*' (1958): 513–526.
7. As so often, the story is ably related in Nilsson, *Quest for AI*; and also Boden, Margaret A. *AI: Its Nature and Future*. Oxford: Oxford University Press, 2016.
8. Minsky, Marvin, and Seymour Papert. *Perceptrons: An Introduction to Computational Geometry*. Cambridge, Mass: MIT Press, 1969.
9. Clark, Liat. 'Google's artificial brain learns to find cat videos', *WIRED UK*, 26 June 2012, https://www.wired.com/2012/06/google-x-neural-network/. For the research, see Le, Quoc V. 'Building high-level features using large scale unsupervised learning', In *2013 IEEE international conference on acoustics, speech and signal processing*, pp. 8595–8598. IEEE, 2013.
10. Mnih, Volodymyr, Koray Kavukcuoglu, David Silver, Andrei A. Rusu, Joel Veness, Marc G. Bellemare, Alex Graves et al. 'Human-level control through deep reinforcement learning', *Nature* 518, no. 7540 (2015): 529–533.

11. See Michel, Arthur Holland. *Eyes in the Sky: The Secret Rise of Gorgon Stare and How it Will Watch Us All*. New York: Houghton Mifflin Harcourt, 2019.
12. Work, Robert O. 'Establishment of an Algorithmic Warfare Cross-Functional Team (ProjectMaven)', US Department of Defense Memorandum, 26 April 2017, https://www.govexec.com/media/gbc/docs/pdfs_edit/establishment_of_the_awcft_project_maven.pdf.
13. Dehaene, Stanislas. *Reading in the Brain: The New Science of How We Read*. New York: Penguin Books, 2010.
14. Martin Ford interviews Bengio, Ng, LeCun, Hinton and many of the other leading lights of modern AI in Ford, Martin. *Architects of Intelligence: The truth about AI from the people building it*. Birmingham: Packt Publishing Ltd, 2018.
15. Marcus, Gary, and Ernest Davis. *Rebooting AI: Building artificial intelligence we can trust*. New York: Pantheon, 2019.
16. Westwick, Peter. *Stealth: The Secret Contest to Invent Invisible Aircraft*. Oxford: Oxford University Press, 2019.
17. See, for an example of current thinking, UK Ministry of Defence. 'Human Machine Teaming', Joint Concept Note 1/18, Development, Concept and Doctrine Centre, May 2018, https://assets.publishing.service.gov.uk/government/uploads/system/uploads/attachment_data/file/709359/20180517-concepts_uk_human_machine_teaming_jcn_1_18.pdf.
18. President's Foreign Intelligence Advisory Board. 'The Soviet "War Scare"' 15 February 1990. Declassified October 2015. https://nsarchive2.gwu.edu/nukevault/ebb533-The-Able-Archer-War-Scare-Declassified-PFIAB-Report-Released/2012-0238-MR.pdf.

i-BATTLE IS JOINED

1. See John Shy, 'Jomini', in Paret, Peter, Gordon Alexander Craig, and Felix Gilbert. *Makers of Modern Strategy: From Machiavelli to the Nuclear Age*. Princeton, N.J.: Princeton University Press, 1986, pp. 143–185.
2. For that argument at greater length, see Payne, Kenneth. 'Artificial intelligence: a revolution in strategic affairs?' *Survival* 60, no. 5 (2018): 7–32.
3. Hoffman, F. G. 'Will war's nature change in the seventh military revolution,' *Parameters*, 47:4, 2017/8, pp. 19–31, at p. 19.

4. For a discussion, see Biddle, Stephen. *Military Power: Explaining Victory and Defeat in Modern Battle*. Princeton, NJ: Princeton University Press, 2010, chapter 5, 'Operation Michael', pp. 78–107.
5. UK Ministry of Defence. 'Land Operations,' Land Warfare Development Centre, Army Doctrine Publication AC 71940, chapter 3, 'Fighting Power', 31 March 2017, https://assets.publishing.service.gov.uk/government/uploads/system/uploads/attachment_data/file/605298/Army_Field_Manual__AFM__A5_Master_ADP_Interactive_Gov_Web.pdf.
6. The literature on the battle is voluminous, but one of the best histories is Overy, R. J. *Battle of Britain: The Myth and Reality*. New York: Norton (W.W) & Co, 2002, and for a first-person account that brings out the human, moral dimension, see Hillary, Richard. *The Last Enemy*. London: Macmillan, 1942.
7. See Metz, Cade. 'The Robots aren't as human as they seem', *The New York Times*, 28 September 2018, https://www.nytimes.com/2018/09/28/technology/robots-human.html.
8. See Sofge, Eric. 'The DARPA Robotics Challenge was a bust', *Popular Science*, 6 July 2015, https://www.popsci.com/darpa-robotics-challenge-was-bust-why-darpa-needs-try-again/.
9. Ernest, Nicholas, David Carroll, Corey Schumacher, Matthew Clark, Kelly Cohen, and Gene Lee. 'Genetic fuzzy based artificial intelligence for unmanned combat aerial vehicle control in simulated air combat missions', *Journal of Defense Management* 6, no. 1 (2016): 2167–0374.
10. DARPA, 'AlphaDogfight trials foreshadow future of human-machine symbiosis,' 26 August 2020, https://www.darpa.mil/news-events/2020-08-26.
11. Osinga, Frans. *Science, Strategy and War: The Strategic Theory of John Boyd*. Abingdon: Routledge, 2007.
12. For a dramatic account of recent state cyber activities, see Sanger, David E. *The Perfect Weapon*. London: Scribe, 2018.
13. Rid, Thomas. *Cyber War Will Not Take Place*. New York: Oxford University Press, 2013.
14. Avgerinos, Thanassis, David Brumley, John Davis, Ryan Goulden, Tyler Nighswander, Alex Rebert, and Ned Williamson. 'The Mayhem cyber reasoning system', *IEEE Security & Privacy* 16, no. 2 (2018): 52–60.

15. Brumley, David. 'Mayhem, the machine that finds software vulnerabilities, then patches them', *IEEE Spectrum*, 29 January 2019, https://spectrum.ieee.org/computing/software/mayhem-the-machine-that-finds-software-vulnerabilities-then-patches-them.
16. Bennhold, Katrinn. 'Merkel Is "Outraged" by Russian Hack but Struggling to Respond', *The New York Times*, 13 May 2020, https://www.nytimes.com/2020/05/13/world/europe/merkel-russia-cyberattack.html.
17. See especially, Beschloss, Michael R. *Kennedy v. Khrushchev: The Crisis Years, 1960–63*. London: Faber and Faber, 1991; and Fursenko, A. A., and Timothy J. Naftali. *One Hell of a Gamble: Khrushchev, Castro, Kennedy, and the Cuban Missile Crisis, 1958–1964*. London: Pimlico, 1999.
18. Elsewhere I argue for the offence, Payne, Kenneth. 'Artificial intelligence: a revolution in strategic affairs?' *Survival* 60, no. 5 (2018): 7–32. Agreement comes from Altmann, Jürgen and Frank Sauer, 'Autonomous Weapon Systems and Strategic Stability', *Survival 59*, no. 5 (2017): 117–142; one argument about defence dominance not yet in print but much discussed among theorists has to do with the vulnerability of AI platforms to electronic warfare. For the shifting balance, see Garfinkel, Ben, and Allan Dafoe. 'How does the offense-defense balance scale?' *Journal of Strategic Studies* 42, no. 6 (2019): 736–763.
19. For a discussion of swarming in warfare, see Scharre, Paul. 'How swarming will change warfare', *Bulletin of the Atomic Scientists* 74, no. 6 (2018): 385–389 and Verbruggen, Maaike, 'The question of swarms control: challenges to ensuring human control over military swarms', *SIPRI Non Proliferation and Disarmament Papers*, no. 65, December 2019, https://www.sipri.org/sites/default/files/2019–12/eunpdc_no_65_031219.pdf. The Perdix swarm is on the US Navy's YouTube channel, at https://youtu.be/bsKbGc9TUHc.
20. Kumar, Vijay. 'Robots that fly and cooperate', TED Talk, 1 March 2012, https://youtu.be/4ErEBkj_3PY.
21. Future of Life Institute, 'Slaughterbots', YouTube, 13 November 2017, https://youtu.be/HipTO_7mUOw.
22. Martin, David. 'Russian hacking proves lethal after Ukranian military

app hijacked', *CBS News*, 22 December 2016, https://www.cbsnews.com/news/russian-hacking-proves-lethal-after-ukrainian-military-app-compromised/.

23. Hern, Alex. 'Fitness tracking app Strava gives away location of secret US army bases', *The Guardian*, 28 January 2018, https://www.theguardian.com/world/2018/jan/28/fitness-tracking-app-gives-away-location-of-secret-us-army-bases.
24. See Kania, Elsa B. 'China's ambitions in artificial intelligence: A challenge to the future of democracy?' *Power 3.0*, 8 August 2018, https://www.power3point0.org/2018/08/08/chinas-ambitions-in-artificial-intelligence-a-challenge-to-the-future-of-democracy/. Democracies and authoritarians sometimes develop exactly the same technology, see Collingridge, John, and Rob Watts, 'Huawei buys stake in UK spy firm Vision Semantics', *The Times*, 19 July 2020, https://www.thetimes.co.uk/article/huawei-buys-stake-in-uk-spy-firm-vision-semantics-65t98vdz0.
25. Thys, Simen, Wiebe Van Ranst, and Toon Goedemé. 'Fooling automated surveillance cameras: adversarial patches to attack person detection', *Proceedings of the IEEE Conference on Computer Vision and Pattern Recognition Workshops*, pp. 49–55. 2019.
26. Ilyas, Andrew, Logan Engstrom, Anish Athalye, and Jessy Lin. 'Black-box adversarial attacks with limited queries and information', *arXiv preprint arXiv:1804.08598* (2018).
27. For the British Army, see chapter 6, 'Mission Command', of Land Warfare Development Centre, 'Land Operations,' Army Doctrine Publication AC 71940, 31 March 2017, https://assets.publishing.service.gov.uk/government/uploads/system/uploads/attachment_data/file/605298/Army_Field_Manual__AFM__A5_Master_ADP_Interactive_Gov_Web.pdf. For a classic American statement, see Krulak, Charles, C. 'The Strategic Corporal: Leadership in the three-block war'. *Marines Magazine* 6, 1999. https://apps.dtic.mil/dtic/tr/fulltext/u2/a399413.pdf.
28. Eady, Yarrow. 'Tesla's deep learning at scale: Using billions of miles to train neural networks', *Towards Data Science*, 7 May 2019, https://towardsdatascience.com/teslas-deep-learning-at-scale-7eed85b235d3.

29. Lye, Harry. 'UK flies 20-drone swarm in major test,' *Airforce Technology*, 28 January 2021, https://www.airforce-technology.com/news/uk-flies-20-drone-swarm-in-major-test/.
30. Cooper, Helene, Ralf Blumenthal and Leslie Kean. '"Wow, what is that?" Navy pilots report Unexplained Flying Objects', *The New York Times*, 26 May 2019, https://www.nytimes.com/2019/05/26/us/politics/ufo-sightings-navy-pilots.html.
31. Harkins, Gina. 'Top Marine General: We need to get comfortable with throwaway equipment,' *Military News*, 2 February 2021, https://www.military.com/daily-news/2021/02/02/top-marine-general-we-need-get-comfortable-throwaway-equipment.html.
32. Insinna, Valerie. 'The US Air Force has built and flown a mysterious full-scale prototype of its future fighter jet,' *Defence News*, 15 September 2020, https://www.defensenews.com/breaking-news/2020/09/15/the-us-air-force-has-built-and-flown-a-mysterious-full-scale-prototype-of-its-future-fighter-jet/.
33. DARPA, 'DARPA initiates design of LongShot Unmanned Air Vehicle,' 8 February 2021, https://www.darpa.mil/news-events/2021-02-08.
34. Lappin, Yaakov. 'IAI announces new Mini Harpy loitering munition', *Janes*, 19 February 2019, https://www.janes.com/defence-news/news-detail/iai-announces-new-mini-harpy-loitering-munition.
35. Athurton, Kelsey D. 'Grenade drone promises an end to cover forever', *Forbes*, 7 May 2020, https://www.forbes.com/sites/kelseyatherton/2020/05/07/grenade-drone-promises-an-end-to-cover-forever/; and Lappin, Yaakov. 'IDF using Uvision mini loitering munition', *Janes*, 16 July 2020, https://www.janes.com/defence-news/news-detail/idf-using-uvision-mini-loitering-munition.
36. Tucker, Patrick. 'Marines are building robotic war balls', *Defense One*, 12 February 2015, https://www.defenseone.com/technology/2015/02/marines-are-building-robotic-war-balls/105258/.
37. For a history, Hennessy, Peter, and James Jinks. *The Silent Deep: The Royal Navy Submarine Service since 1945*. London: Penguin Books, 2016.
38. Kaplan, Fred. *The Bomb: Presidents, Generals, and the Secret History of the Nuclear War*. New York: Simon & Schuster, 2020.

39. On some of the challenges, see Brose, Christian. *The Kill Chain: Defending America in the Future of High-Tech Warfare*. New York: Hachette, 2020.
40. For more, see Farrell, Theo and Kenneth Payne, 'Artificial Intelligence and strategic culture,' in Kartchner, Kerry, et al. (ed). *The Routledge Handbook of Strategic Culture*. Abingdon: Routledge, 2021/2 (forthcoming).
41. See Dunbar, Robin I.M. 'Neocortex size and group size in primates: A test of the hypothesis', *Journal of Human Evolution* 28, no. 3 (1995): 287–296; and Payne, Kenneth. *The Psychology of Modern Conflict: Evolutionary Theory, Human Nature and a Liberal Approach to War*. Basingstoke: Palgrave Macmillan, 2015.
42. For a discussion see King, Anthony. *The Combat Soldier: Infantry Tactics and Cohesion in the Twentieth and Twenty-first Centuries*. Oxford: Oxford University Press, 2013.
43. https://arxiv.org/pdf/2007.08794.pdf.
44. NASA, 'Borg computer collective designs NASA space antenna', 16 February 2006, https://www.nasa.gov/centers/ames/research/exploringtheuniverse/borg.html.
45. Surdu, John R., and Kevin Kittka. 'Deep Green: Commander's tool for COA's Concept', In *Proceedings of Computing, Communications and Control Technology Conference*, vol. 29. 2008.
46. Chung, Timothy. 'Offensive swarm enabled tactics', *DARPA*, undated release, https://www.darpa.mil/program/offensive-swarm-enabled-tactics.
47. Lye, Harry. 'RAF to launch swarming drone squadron in April', *Airforce Technology*, 13 January 2020, https://www.airforce-technology.com/news/raf-swarming-drones/.
48. Mizokami, Kyle. 'This robot tank is a glimpse at the future of armoured warfare', *Popular Mechanics*, 23 June 2020, https://www.popularmechanics.com/military/weapons/a32947926/type-x-robot-combat-vehicle/.
49. Franklin-Wallace, Oliver. 'If we are living in a simulation, this UK start up probably built it', *WIRED*, 11 May 2017, https://www.wired.co.uk/article/improbable-quest-to-build-the-matrix; and for the defence con-

nection, Cook, James. 'Ministry of Defence pays improbable £8.3m for wargaming software', *The Telegraph*, 15 June 2020, https://www.telegraph.co.uk/technology/2020/06/15/ministry-defence-pays-improbable-83m-wargaming-software/.
50. Keller, Jared. 'US special operations forces are testing a "guaranteed hit" smart rifle system in Syria', *Task and Purpose*, 5 June 2020, https://taskandpurpose.com/military-tech/special-operations-smart-shooter-syria.
51. Mizokami, Kyle. 'Become a super sniper: DARPA is turning 0.50 caliber bullets into guided rounds', *The National Interest*, 28 July 2018, https://nationalinterest.org/blog/buzz/become-super-sniper-darpa-turning-50-caliber-bullets-guided-rounds-27101.
52. Stewart, Jack. 'Israel's self-flying "Cormorant" whisks soldiers to safety', *WIRED*, 26 May 2018, https://www.wired.com/story/tactical-robotics-cormorant-autonomous/.
53. Wong, Julia Carrie. 'The FBI and Apple are facing off over an iPhone again. What's going on?' *The Guardian*, 15 January 2020, https://www.theguardian.com/us-news/2020/jan/14/fbi-apple-faceoff-iphone-florida-shooting.
54. Jensen, Benjamin and John Paschkewitz. 'Mosaic warfare: Small and scalable are beautiful', *War on the Rocks*, 23 December 2019, https://warontherocks.com/2019/12/mosaic-warfare-small-and-scalable-are-beautiful/.

MINDS MAKE STRATEGY

1. Madrigal, Alexis C. 'The computer that predicted the US would win the Cold War', *The Atlantic*, 5 October 2017, https://www.theatlantic.com/technology/archive/2017/10/the-computer-that-predicted-the-us-would-win-the-vietnam-war/542046/.
2. Jill Lepore has described early attempts to use computers for this sort of predictive task, including during the Vietnam war. As you'd suspect, the efforts were stymied by limited data, and limited computing power. Lepore, Jill. *If Then: How One Data Company Invented the Future*. London: John Murray, 2020.
3. Ellsberg, Daniel. *The Doomsday Machine: Confessions of a Nuclear War Planner*. London, 2018.

4. Ellsberg, Daniel. *Secrets: A Memoir of Vietnam and the Pentagon Papers*. London: Penguin Books, 2003.
5. Pawlyk, Oriana. 'The Air Force risks a drone pilot shortage, GAO finds', *Military.com*, 26 June 2020, https://www.military.com/daily-news/2020/06/26/air-force-risks-drone-pilot-shortage-gao-finds.html; Fisher, Lucy. 'Stress of killing from afar creates shortage of MoD drone operators', *The Times*, 13 January 2020, https://www.thetimes.co.uk/article/stress-of-killing-from-afar-creates-shortage-of-mod-drone-operators-0fnm36r6t.
6. Allen, Gregory C. 'Project Maven brings AI to the fight against ISIS,' *Bulletin of the Atomic Scientists*, 21 December 2017, https://thebulletin.org/2017/12/project-maven-brings-ai-to-the-fight-against-isis/.
7. Lohr, Steve. 'How do you vote? 50 million Google images give a clue', *The New York Times*, 31 December 2017, https://www.nytimes.com/2017/12/31/technology/google-images-voters.html.
8. Vigen, Tyler. *Spurious Correlations*. New York: Hachette Books, 2015.
9. See Crawford, Kate et al. *AI Now 2019 Report*. New York: AI Now Institute, 2019, Section 2.5 Facial/affect recognition, pp. 50–52, https://ainowinstitute.org/AI_Now_2019_Report.pdf#page=50; and on the general problem, Barrett, Lisa Feldman et al. 'Emotional expressions reconsidered: Challenges to inferring emotion from human facial movements', *Psychological Science in the Public Interest* 20, no. 1 (2019): 1–68.
10. Dressel, Julia, and Hany Farid. 'The accuracy, fairness, and limits of predicting recidivism', *Science Advances* 4, no. 1 (2018): eaao5580. Subsequent research found AI outperformed humans but was still less than 70 percent accurate: Lin, Z., J. Jung, S. Goel, and J. Skeem. 'The limits of human predictions of recidivism', *Science Advances*, 6(7) (2020), https://doi.org/10.1126/sciadv.aaz0652.
11. Dastin, Jeffrey. 'Amazon scraps secret AI recruiting tool that showed bias against women', *Reuters*, 10 October 2018, https://www.reuters.com/article/us-amazon-com-jobs-automation-insight/amazon-scraps-secret-ai-recruiting-tool-that-showed-bias-against-women-idUSKCN1MK08G.
12. Singer, Natasha. 'Amazon is pushing facial technology that a study says could be biased', *The New York Times*, 24 January 2019, https://

www.nytimes.com/2019/01/24/technology/amazon-facial-technology-study.html.

13. For the original research, Wang, Yilun, and Michal Kosinski. 'Deep neural networks are more accurate than humans at detecting sexual orientation from facial images', *Journal of Personality and Social Psychology* 114, no. 2 (2018): 246. For the critique, Leuner, John. 'A Replication Study: Machine Learning Models are capable of predicting sexual orientation from facial images', *arXiv preprint:1902.10739* (2019).
14. Johnson, Lyndon B. *The Vantage Point: Perspectives of the Presidency 1963–1969*. New York: Holt, Reinhart and Winston, 1971, p. 365.
15. Kennedy's reflection came in the introduction he wrote to Sorensen, Theodore C. *Decision-making in the White House: The Olive Branch or the Arrows*. New York: Columbia University Press, 1963, p. xxxi.
16. See my forthcoming article, 'Empathy and enmity: Kennedy and Khrushchev's dance of minds'. The classic study is White, Ralph K. *Fearful Warriors: A Psychological Profile of U.S.-Soviet Relations*. London: Collier Macmillan, 1984.
17. BBC World News Impact, 'Interview with Mark Sagar and Baby X', 4 October 2013, https://youtu.be/XBsl3IIlB8VE. For the research, Sagar, Mark , Mike Seymour, and Annette Henderson. 'Creating connection with autonomous facial animation', *Communications of the ACM*. 59. 82–91. 10.1145/2950041 (2016).
18. Knapton, Sarah. 'Robot that can feel pain invented by scientist', *The Telegraph*, 22 February 2020, https://www.telegraph.co.uk/science/2020/02/22/watch-robot-can-feel-pain-invented-scientists/.
19. For example, see Smith, Adam. 'Cognitive empathy and emotional empathy in human behavior and evolution', *The Psychological Record* 56, no. 1 (2006): 3–21.
20. Abadi, Martín, and David G. Andersen. 'Learning to protect communications with adversarial neural cryptography', *arXiv preprint arXiv:1610.06918* (2016). https://arxiv.org/pdf/1610.06918v1.pdf.
21. The ideas here draw on both Schelling's books. Schelling, Thomas C. *The Strategy of Conflict*. New York: Oxford University Press, 1963 and Schelling, Thomas C. *Arms and Influence*. 2008. New Haven: Yale University Press, 1966.

22. See Kaplan, Fred M. *The Wizards of Armageddon*. Stanford, Calif.: Stanford University Press, 1991.
23. On the SIOP see Kaplan, Fred. *The Bomb: Presidents, Generals, and the Secret History of the Nuclear War* New York: Simon & Schuster, 2020; also Freedman, Lawrence, and Michaels, Jeffrey. *The Evolution of Nuclear Strategy*. Fourth ed. London: Palgrave Macmillan, 2019, especially chapter 17, pp. 259–278.
24. Blight, James G. *The Shattered Crystal Ball: Fear and Learning in the Cuban Missile Crisis*. Savage, Md: Rowman & Littlefield, 1990.
25. Markwica, Robin. *Emotional Choices: How the Logic of Affect Shapes Coercive Diplomacy*. Oxford: Oxford University Press, 2018.
26. Clausewitz, Carl Von. *On War*. London: David Campbell, 1993, p. 125.
27. Brown, Noam, and Tuomas Sandholm. 'Superhuman AI for heads-up no-limit poker: Libratus beats top professionals', *Science* 359, no. 6374 (2018): 418–424.
28. For the story of Libratus, see Metz, Cade, 'Inside Libratus, the poker AI that out-bluffed the best humans', *WIRED*, 1 February, 2017, https://www.wired.com/2017/02/libratus/.
29. Heaven, Douglas. 'No limit: AI poker bot is first to beat professionals at multiplayer game', *Nature* 571, no. 7765 (2019): 307–309.
30. Senator, Ted, 'Serial Interactions in Imperfect Information Games Applied to Complex Military Decision Making (SI3-CMD)', DARPA program information, undated, https://www.darpa.mil/program/serial-interactions-in-imperfect-information-games-applied-to-complex-military-decision-making.

THE ART OF i-WAR

1. Gladwell, Malcolm. *Outliers: The Story of Success*. London: Allen Lane, 2008.
2. To see the experiment, Sandlin, Destin. 'The backward bicycle brain', *Smarter Every Day*, 24 April 2015, https://youtu.be/MFzDaBzBlL0.
3. Boden, Margaret A. *Creativity and Art: Three Roads to Surprise*. Oxford: Oxford University Press, 2010.
4. Kasparov describes his creative process himself, in his own book about machine creativity—Kasparov, Garry K. and Mig Greengard. *Deep*

Thinking: Where Machine Intelligence Ends and Human Creativity Begins. London: John Murray, 2018.

5. For the research see, Silver, David, Aja Huang, Chris J. Maddison, Arthur Guez, Laurent Sifre, George Van Den Driessche, Julian Schrittwieser et al. 'Mastering the game of Go with deep neural networks and tree search', *Nature* 529, no. 7587 (2016): 484–489. For the drama, see Greg Koh's 2017 documentary, *AlphaGo*.
6. Schrittwieser, Julian, Ioannis Antonoglou, Thomas Hubert, Karen Simonyan, Laurent Sifre, Simon Schmitt, Arthur Guez et al. 'Mastering Atari, go, chess and shogi by planning with a learned model', *arXiv preprint arXiv:1911.08265* (2019).
7. Creswell, Antonia, Kyriacos Nikiforou, Oriol Vinyals, Andre Saraiva, Rishabh Kabra, Loic Matthey, Chris Burgess et al. 'AlignNet: Unsupervised Entity Alignment', *arXiv preprint arXiv:2007.08973* (2020).
8. DARPA, 'AI Next Campaign', undated release, https://www.darpa.mil/work-with-us/ai-next-campaign.
9. Maslow, Abraham. 'A theory of human motivation', *Psychological Review* 50, no. 4 (1943): 370–396.

HUMAN-MACHINE TEAMS

1. Jaderberg, Max, Wojciech M. Czarnecki, Iain Dunning, Luke Marris, Guy Lever, Antonio Garcia Castaneda, Charles Beattie et al. 'Human-level performance in 3D multiplayer games with population-based reinforcement learning', *Science* 364, no. 6443 (2019): 859–865.
2. Berner, Christopher, Greg Brockman, Brooke Chan, Vicki Cheung, Przemysław Dębiak, Christy Dennison, David Farhi et al. 'Dota 2 with large scale deep reinforcement learning', *arXiv preprint arXiv:1912.06680* (2019).
3. See OpenAI, 'OpenAI Five defeats Dota 2 World Champions', 15 April 2019, https://openai.com/blog/openai-five-defeats-dota-2-world-champions/.
4. Cohn, Gabe. 'AI art at Christies sells for $432,500', *The New York Times*, 25 October 2018, https://www.nytimes.com/2018/10/25/arts/design/ai-art-sold-christies.html.

5. Elgammal, Ahmed, Bingchen Liu, Mohamed Elhoseiny, and Marian Mazzone. 'Can: Creative adversarial networks, generating "art" by learning about styles and deviating from style norms', *arXiv preprint arXiv:1706.07068* (2017).
6. Seabrook, John. 'The next word: Where will predictive text take us?' *The New Yorker*, 14 October 2019 https://www.newyorker.com/magazine/2019/10/14/can-a-machine-learn-to-write-for-the-new-yorker.
7. Stahl, Jennifer. 'Could Google be the world's next great choreographer?' *Dance Magazine*, 11 January 2019, https://www.dancemagazine.com/is-google-the-worlds-next-great-choreographer-2625652667.html.
8. Kelly, Mary Louise. 'One key, many notes: Ólafur Arnalds' piano rig fuses technology and musicality', *NPR All Things Considered*, 19 July 2018, https://www.npr.org/2018/07/19/630111211/one-key-many-notes-lafur-arnalds-piano-rig-fuses-technology-and-musicality?t=1595412287661.
9. Marshall, Alex. 'A robot walks into a bar. But can it do comedy?' *The New York Times*, 8 August 2018, https://www.nytimes.com/2018/08/08/arts/ai-comedy-artificial-intelligence-piotr-mirowski.html; and Improbotics: An Improvised Theatre Experiment, https://improbotics.org.
10. Neubauer, Catherine, Gregory Gremillion, Brandon S. Perelman, Claire La Fleur, Jason S. Metcalfe, and Kristin E. Schaefer. 'Analysis of facial expressions explain affective state and trust-based decisions during interaction with autonomy.' In *International Conference on Intelligent Human Systems Integration*, pp. 999–1006.
11. Wong, Yuna Huh, John Yurchak, Robert W. Button, Aaron Frank, Burgess Laird, Osonde A. Osoba, Randall Steeb, Benjamin N. Harris, and Sebastian Joon Bae, *Deterrence in the Age of Thinking Machines*. Santa Monica, CA: RAND Corporation, 2020. https://www.rand.org/pubs/research_reports/RR2797.html.
12. Russon, Mary-Ann. 'The robot that watches as you cross the road', *BBC News*, 28 August 2018, https://www.bbc.co.uk/news/business-45330451.
13. Schelling, Thomas. *Arms and Influence* New Haven: Yale University Press, 1966, p. 116.

14. Schelling, op cit, p. 118.
15. Osborn, Kris, 'Future of war will be "hyperactive battlefields": US Army General,' *The National Interest*, 30 January 2021, https://nationalinterest.org/blog/buzz/future-war-will-be-'hyperactive-battlefields'-us-army-general-177371.
16. Russell, Stuart J. *Human Compatible: Artificial Intelligence and the Problem of Control*. London: Allen Lane, 2019.
17. Bostrom, Nick. *Superintelligence: Paths, Dangers, Strategies*. Oxford: Oxford University Press, 2016.
18. On the Good Judgment Project, see Tetlock, Philip E., Barbara A. Mellers, and J. Peter Scoblic. 'Bringing probability judgments into policy debates via forecasting tournaments', *Science* 355, no. 6324 (2017): 481–483. The sample questions are taken from the HFC Challenge 'pre-season', https://www.gjopen.com/challenges/19-coming-soon-hfc-challenge?status=closed.
19. On SAGE see Morstatter, Fred, Aram Galstyan, Gleb Satyukov, Daniel Benjamin, Andres Abeliuk, Mehrnoosh Mirtaheri, K. S. M. Hossain et al. 'SAGE: a hybrid geopolitical event forecasting system', In *Proceedings of the 28th International Joint Conference on Artificial Intelligence*, pp. 6557–6559. AAAI Press, 2019.
20. Cole, August. 'HOLESHOT', *The RUSI Journal* 164, no. 5–6 (2019): 12–16.
21. For an overview of recent developments, see Poldrack, Russell A. *The New Mind Readers: What Neuroimaging Can and Cannot Reveal about Our Thoughts*. Princeton, N.J.: Princeton University Press, 2018.
22. Collinger, Jennifer L., Brian Wodlinger, John E. Downey, Wei Wang, Elizabeth C. Tyler-Kabara, Douglas J. Weber, Angus J. C. McMorland, Meel Velliste, Michael L. Boninger, and Andrew B. Schwartz. 'High-performance neuroprosthetic control by an individual with tetraplegia', *The Lancet* 381, no. 9866 (2013): 557–564.
23. Stockton, Nick. 'Woman controls a fighter jet sim using only her mind', *WIRED*, 5 March 2015, https://www.wired.com/2015/03/woman-controls-fighter-jet-sim-using-mind/.
24. Ortiz-Catalan, Max, Enzo Mastinu, Paolo Sassu, Oskar Aszmann, Rickard Brånemark. 'Self-Contained Neuromusculoskeletal Arm Prostheses', *New England Journal of Medicine* 382, no. 18 (2020): 1732.

25. Twilley, Nicola. 'Seeing with your tongue,' *The New York Times*, 8 May 2017, https://www.newyorker.com/magazine/2017/05/15/seeing-with-your-tongue.
26. Nagel, Thomas. 'What Is It Like to Be a Bat?' *The Philosophical Review* 83, no. 4 (1974): 435–50.
27. David Eagleman, 'Can we create new senses for humans', *TED talks*, March 2015, https://www.ted.com/talks/david_eagleman_can_we_create_new_senses_for_humans?utm_campaign=tedspread&utm_medium=referral&utm_source=tedcomshare.
28. Pais-Vieira, Miguel, Mikhail Lebedev, Carolina Kunicki, Jing Wang, and Miguel A. L. Nicolelis. 'A brain-to-brain interface for real-time sharing of sensorimotor information', *Scientific Reports* 3 (2013): 1319. Zhang, Shaomin, Sheng Yuan, Lipeng Huang, Xiaoxiang Zheng, Zhaohui Wu, Kedi Xu, and Gang Pan. 'Human mind control of rat cyborg's continuous locomotion with wireless brain-to-brain interface', *Scientific Reports* 9, no. 1 (2019): 1–12; Jiang, Linxing, Andrea Stocco, Darby M. Losey, Justin A. Abernethy, Chantel S. Prat, and Rajesh P. N. Rao. 'BrainNet: a multi-person brain-to-brain interface for direct collaboration between brains', *Scientific Reports* 9, no. 1 (2019): 1–11.
29. See New Scientist, 'Cyborg insects with wings controlled by humans', March 2008, https://youtu.be/dSCLBG9KeX4; and for the research Sato, Hirotaka, Christopher W. Berry, Brendan E. Casey, Gabriel Lavella, Ying Yao, John M. VandenBrooks, and Michel M. Maharbiz. 'A cyborg beetle: insect flight control through an implantable, tether-less microsystem', In *2008 IEEE 21st International Conference on Micro Electro Mechanical Systems*, pp. 164–167. IEEE, 2008.
30. Damasio, Antonio R. *Descartes' Error: Emotion, Reason and the Human Brain*. London: Vintage, 2006.
31. Slovic, Paul, Melissa L. Finucane, Ellen Peters, and Donald G. MacGregor. 'Risk as analysis and risk as feelings: Some thoughts about affect, reason, risk, and rationality', *Risk Analysis: An International Journal* 24, no. 2 (2004): 311–322.
32. See LeDoux, Joseph E. *Anxious: The Modern Mind in the Age of Anxiety*. London: Oneworld, 2015.
33. Lin, Dayu, Maureen P. Boyle, Piotr Dollar, Hyosang Lee, E. S. Lein,

Pietro Perona, and David J. Anderson. 'Functional identification of an aggression locus in the mouse hypothalamus', *Nature* 470, no. 7333 (2011): 221–226.

34. Anpilov, Sergey, Yair Shemesh, Noa Eren, Hala Harony-Nicolas, Asaf Benjamin, Julien Dine, Vinícius E. M. Oliveira et al. 'Wireless Optogenetic Stimulation of Oxytocin Neurons in a Semi-natural Setup Dynamically Elevates Both Pro-social and Agonistic Behaviors', *Neuron*, in press, (2020).
35. See Mitchell, Kevin J. *Innate: How the Wiring of Our Brains Shapes Who We Are*. Princeton, NJ: Princeton University Press, 2018.
36. See Plomin, Robert. *Blueprint: How DNA Makes Us Who We Are*. London: Penguin, 2019.
37. Ledford, Heidi. 'CRISPR treatment inserted directly into the body for first time', *Nature* 579, no. 7798 (2020): 185–185.
38. Regalado, Antonio, 'Exclusive: Chinese scientists are creating CRISPR babies', *MIT Technology Review*, 25 November 2018, https://www.technologyreview.com/2018/11/25/138962/exclusive-chinese-scientists-are-creating-crispr-babies/.
39. Fu, Rui, Dawei Yu, Jilong Ren, Chongyang Li, Jing Wang, Guihai Feng, Xuepeng Wang et al. 'Domesticated cynomolgus monkey embryonic stem cells allow the generation of neonatal interspecies chimeric pigs', *Protein & Cell* 11, no. 2 (2020): 97–107.
40. Lei Shi, Xin Luo, Jin Jiang, Yongchang Chen, Cirong Liu, Ting Hu, Min Li, Qiang Lin, Yanjiao Li, Jun Huang, Hong Wang, Yuyu Niu, Yundi Shi, Martin Styner, Jianhong Wang, Yi Lu, Xuejin Sun, Hualin Yu, Weizhi Ji, Bing Su, 'Transgenic rhesus monkeys carrying the human *MCPH1* gene copies show human-like neoteny of brain development', *National Science Review* 6, no. 3 (May 2019): 480–493.
41. Martin, C., R. Bhui, P. Bossaerts et al. 'Chimpanzee choice rates in competitive games match equilibrium game theory predictions'. *Scientific Reports* 4, 5182 (2015).
42. Inoue, Sana, and Tetsuro Matsuzawa. 'Working memory of numerals in chimpanzees', *Current Biology* 17, no. 23 (2007): R1004-R1005.
43. Ryan, Mick and Therese Keane, 'Biotechnology and human augmentation: implications for National Security practitioners,' *The Strategy*

Bridge, 5 Feburary 2019, https://thestrategybridge.org/the-bridge/2019/2/5/biotechnology-and-human-augmentation-issues-for-national-security-practitioners.

44. FR24 News, 'France gives [ethical approval] to "augmented soldiers"', 12 December 2020, https://www.fr24news.com/a/2020/12/france-gives-a-sign-of-ethics-to-augmented-soldiers.html.

RULES FOR WARBOTS

1. Perrow, Charles. *Normal Accidents: Living with High-risk Technologies*. New York: Basic, 1984.
2. Maas, Matthijs M. 'How Viable Is International Arms Control for Military Artificial Intelligence? Three Lessons from Nuclear Weapons', *Contemporary Security Policy* 40, no. 3 (2019): 285–311. See also for an extended discussion about whether normal accident theory or high reliability organisations typifies AI, Scharre, Paul, 'Autonomous weapons and stability', unpublished PhD thesis, King's College London, 2020.
3. On the evolution of morality, see Boehm, Christopher. *Moral Origins: The Evolution of Virtue, Altruism, and Shame*. New York: Basic Books, 2012; and Wright, Robert. *The Moral Animal: Evolutionary Psychology and Everyday Life*. London: Little, Brown, 1995.
4. Geertz, Clifford, and Robert Darnton. *The Interpretation of Cultures: Selected Essays*. New York: Basic Books, 2017.
5. Gottschall, Jonathan. *The Storytelling Animal: How Stories Make Us Human*. Boston Mass.: Houghton Mifflin Harcourt, 2012.
6. Trivers, Robert L. 'The evolution of reciprocal altruism', *The Quarterly Review of Biology* 46, no. 1 (1971): 35–57.
7. Rawls, John. *A Theory of Justice*. Boston, M.A.: Harvard University Press, 2009.
8. Mill, John Stuart, and Gertrude Himmelfarb. *On Liberty*. Harmondsworth: Penguin, 1985.
9. Bull, Hedley. *The Anarchical Society: A Study of Order in World Politics*. London: Macmillan, 1977.
10. Wrangham, Richard W. *The Goodness Paradox: The Strange Relationship between Virtue and Violence in Human Evolution*. New York: Vintage Books, 2019.

11. The landmark modern text is Walzer, Michael. *Just and Unjust Wars: A Moral Argument with Historical Illustrations*. New York: Basic Books, 1977. A modern overview is Whetham, David, and Fisher, David. *Ethics, Law and Military Operations*. Basingstoke: Palgrave Macmillan, 2011.
12. Sini, Rozina. 'Does Saudi robot citizen have more rights than women?' *BBC News*, 26 October 2017, https://www.bbc.co.uk/news/blogs-trending-41761856.
13. Greene, Joshua D. *Moral Tribes: Emotion, Reason, and the Gap between Us and Them*. New York: Penguin Press, 2013.
14. See DARPA, 'Towards a high resolution implantable neural interface,' 10 July 2017, https://www.darpa.mil/news-events/2017-07-10.
15. Gregory, Andy. 'Coronavirus: Former MI6 head claims pandemic "started as accident" in Chinese laboratory', *The Independent*, 4 June 2020, https://www.independent.co.uk/news/world/asia/coronavirus-chinese-laboratory-wuhan-mi6-richard-dearlove-conspiracy-theories-a9547851.html.
16. Strassler, Robert B, and Richard Crawley. *The Landmark Thucydides: A Comprehensive Guide to the Peloponnesian War*. London: Free Press, 1996; and for the modern spinoff, Allison, Graham T. *Destined for War: Can America and China Escape Thucydides' Trap?* London: Scribe, 2017.
17. Two of the most perceptive scholars of China's military AI and the prospects of an AI arms race are Elsa Kania and Jim Johnson. See Kania, Elsa B. 'The pursuit of AI is more than an arms race', *Defense One*, 19 April 2018, https://www.defenseone.com/ideas/2018/04/pursuit-ai-more-arms-race/147579/; and Johnson, James. *The AI Revolution & the Future of Warfare: Great Power Competition & Strategic Stability*. Manchester: Manchester University Press, forthcoming 2021.
18. Wohlstetter, Albert. 'The delicate balance of terror', *Foreign Affairs* 37 (1958): 211.
19. UK Ministry of Defence, 'JDP 0–30.2: Unmanned Aircraft Systems', Development, Doctrine and Concepts Centre, August 2017, https://assets.publishing.service.gov.uk/government/uploads/system/uploads/attachment_data/file/673940/doctrine_uk_uas_jdp_0_30_2.pdf, p. 13.
20. See National Security Commission on Artificial Intelligence, 'Draft

Final Report,' January 2021, https://www.nscai.gov/wp-content/uploads/2021/01/NSCAI-Draft-Final-Report-1.19.21.pdf.

21. Felter, Joseph H., and Jacob N. Shapiro. 'Limiting Civilian Casualties as Part of a Winning Strategy: The Case of Courageous Restraint', *Daedalus* 146, no. 1 (2017): 44–58.
22. See Bryson, J.J., M. E. Diamantis, and T. D. Grant. 'Of, for, and by the people: the legal lacuna of synthetic persons', *Artificial Intelligence and Law* 25, (2017): 273–291.
23. Sorensen, Theodore C. *Kennedy*. New York: Harper Perennial, 2009, p. 527.

BIBLIOGRAPHY

'A proposal for the Dartmouth summer research project on Artificial Intelligence', 31 August 1955, http://www-formal.stanford.edu/jmc/history/dartmouth/dartmouth.html.

Abadi, Martín, and David G. Andersen. 'Learning to protect communications with adversarial neural cryptography', *arXiv preprint arXiv:1610.06 918* (2016). https://arxiv.org/pdf/1610.06918v1.pdf.

Adams, Douglas. *The Hitchhiker's Guide to the Galaxy*. London: William Heinemann, 2014.

Allen, Gregory C. 'Project Maven brings AI to the fight against ISIS,' *Bulletin of the Atomic Scientists*, 21 December 2017, https://thebulletin.org/2017/12/project-maven-brings-ai-to-the-fight-against-isis/.

Allison, Graham T. *Destined for War: Can America and China Escape Thucydides' Trap?* London: Scribe, 2017.

Altmann, Jürgen and Frank Sauer, 'Autonomous Weapon Systems and Strategic Stability', *Survival* 59, no. 5 (2017): 117–142.

Anpilov, Sergey, Yair Shemesh, Noa Eren, Hala Harony-Nicolas, Asaf Benjamin, Julien Dine, Vinícius E. M. Oliveira et al. 'Wireless Optogenetic Stimulation of Oxytocin Neurons in a Semi-natural Setup Dynamically Elevates Both Pro-social and Agonistic Behaviors', *Neuron*, in press, (2020).

Ashby, Ross, and Paul R. Stein. 'Design for a Brain', *PhT* 7, no. 4 (1954): 24.

Asimov, Isaac. *I, Robot*. London: Harper Voyager, 2013.

BIBLIOGRAPHY

Athurton, Kelsey D. 'Grenade drone promises an end to cover forever', *Forbes*, 7 May 2020, https://www.forbes.com/sites/kelseyatherton/2020/05/07/grenade-drone-promises-an-end-to-cover-forever/.

Avgerinos, Thanassis, David Brumley, John Davis, Ryan Goulden, Tyler Nighswander, Alex Rebert, and Ned Williamson. 'The Mayhem cyber reasoning system', *IEEE Security & Privacy* 16, no. 2 (2018): 52–60.

Baard, Mark. 'AI founder blasts modern research', *WIRED*, 13 May 2003, https://www.wired.com/2003/05/ai-founder-blasts-modern-research/.

Banks, Iain. *The Player of Games*. London: Orbit, 2005.

Barrett, Lisa Feldman et al. 'Emotional expressions reconsidered: Challenges to inferring emotion from human facial movements', *Psychological Science in the Public Interest* 20, no. 1 (2019): 1–68.

BBC World News Impact, 'Interview with Mark Sagar and Baby X', 4 October 2013, https://youtu.be/XBsl3HlB8VE.

Beer, Randall D. 'The dynamics of adaptive behavior: A research program', *Robotics and Autonomous Systems* 20, no. 2–4 (1997): 257–289.

Bennhold, Katrinn. 'Merkel Is "Outraged" by Russian Hack but Struggling to Respond', *The New York Times*, 13 May 2020, https://www.nytimes.com/2020/05/13/world/europe/merkel-russia-cyberattack.html.

Berner, Christopher, Greg Brockman, Brooke Chan, Vicki Cheung, Przemysław Dębiak, Christy Dennison, David Farhi et al. 'Dota 2 with large scale deep reinforcement learning', *arXiv preprint arXiv:1912.06680* (2019).

Beschloss, Michael R. *Kennedy v. Khrushchev: The Crisis Years, 1960–63*. London: Faber and Faber, 1991.

Biddle, Stephen. *Military Power: Explaining Victory and Defeat in Modern Battle*. Princeton, NJ: Princeton University Press, 2010.

Blight, James G. *The Shattered Crystal Ball: Fear and Learning in the Cuban Missile Crisis*. Savage, Md: Rowman & Littlefield, 1990.

Boden, Margaret A. *Creativity and Art: Three Roads to Surprise*. Oxford: Oxford University Press, 2010.

Boden, Margaret A. *AI: Its Nature and Future*. Oxford: Oxford University Press, 2016.

Boehm, Christopher. *Moral Origins: The Evolution of Virtue, Altruism, and Shame*. New York: Basic Books, 2012.

BIBLIOGRAPHY

Bostrom, Nick. *Superintelligence: Paths, Dangers, Strategies*. Oxford: Oxford University Press, 2016.

Brooks, Rodney A. 'Intelligence without representation', *Artificial Intelligence* 47, no. 1–3 (1991): 139–159.

Brose, Christian. *The Kill Chain: Defending America in the Future of High-Tech Warfare*. New York: Hachette, 2020.

Brown, Noam, and Tuomas Sandholm. 'Superhuman AI for heads-up no-limit poker: Libratus beats top professionals', *Science* 359, no. 6374 (2018): 418–424.

Brumley, David. 'Mayhem, the machine that finds software vulnerabilities, then patches them', *IEEE Spectrum*, 29 January 2019, https://spectrum.ieee.org/computing/software/mayhem-the-machine-that-finds-software-vulnerabilities-then-patches-them.

Bryson, J.J., M. E. Diamantis, and T. D. Grant. 'Of, for, and by the people: the legal lacuna of synthetic persons', *Artificial Intelligence and Law* 25, (2017): 273–291.

Bull, Hedley. *The Anarchical Society: A Study of Order in World Politics*. London: Macmillan, 1977.

Cellan-Jones, Rory. 'Stephen Hawking warns artificial intelligence could end mankind', *BBC News*, 2 December 2014.

Chomsky, Noam. *Aspects of the Theory of Syntax*. Cambridge, Mass: MIT, 1965.

Chung, Timothy. 'Offensive swarm enabled tactics', *DARPA*, undated release, https://www.darpa.mil/program/offensive-swarm-enabled-tactics.

Clark, Liat. 'Google's artificial brain learns to find cat videos', *WIRED UK*, 26 June 2012, https://www.wired.com/2012/06/google-x-neural-network/.

Clausewitz, Carl Von. *On War*. London: David Campbell, 1993.

Cobb, Matthew. *The Idea of the Brain: A History*. London: Profile Books, 2020.

Cohn, Gabe. 'AI art at Christies sells for $432,500', *The New York Times*, 25 October 2018, https://www.nytimes.com/2018/10/25/arts/design/ai-art-sold-christies.html.

Cole, August. 'HOLESHOT', *The RUSI Journal* 164, no. 5–6 (2019): 12–16.

Collinger, Jennifer L., Brian Wodlinger, John E. Downey, Wei Wang, Elizabeth C. Tyler-Kabara, Douglas J. Weber, Angus J. C. McMorland, Meel Velliste, Michael L. Boninger, and Andrew B. Schwartz. 'High-performance neuroprosthetic control by an individual with tetraplegia', *The Lancet* 381, no. 9866 (2013): 557–564.

Collingridge, John and Rob Watts. 'Huawei buys stake in UK spy firm Vision Semantics', *The Times*, 19 July 2020, https://www.thetimes.co.uk/article/huawei-buys-stake-in-uk-spy-firm-vision-semantics-65t98vdz0.

Cook, James. 'Ministry of Defence pays improbable £8.3m for wargaming software', *The Telegraph*, 15 June 2020, https://www.telegraph.co.uk/technology/2020/06/15/ministry-defence-pays-improbable-83m-wargaming-software/.

Cooper, Helene, Ralf Blumenthal and Leslie Kean, '"Wow, what is that?" Navy pilots report Unexplained Flying Objects', *The New York Times*, 26 May 2019, https://www.nytimes.com/2019/05/26/us/politics/ufo-sightings-navy-pilots.html.

Crawford, Kate, et al. *AI Now 2019 Report*. New York: AI Now Institute, 2019, Section 2.5 Facial/affect recognition, pp. 50–52, https://ainowinstitute.org/AI_Now_2019_Report.pdf#page=50.

Creswell, Antonia, Kyriacos Nikiforou, Oriol Vinyals, Andre Saraiva, Rishabh Kabra, Loic Matthey, Chris Burgess et al. 'AlignNet: Unsupervised Entity Alignment', *arXiv preprint arXiv:2007.08973* (2020).

Damasio, Antonio R. *The Feeling of What Happens: Body and Emotion in the Making of Consciousness*. London: W. Heinemann, 2000.

Damasio, Antonio R. *Descartes' Error: Emotion, Reason and the Human Brain*. London: Vintage, 2006.

DARPA, 'AI Next Campaign', undated release, https://www.darpa.mil/work-with-us/ai-next-campaign.

DARPA, 'Towards a high resolution implantable neural interface,' 10 July 2017, https://www.darpa.mil/news-events/2017-0-10.

DARPA, 'AlphaDogfight trials foreshadow future of human-machine symbiosis,' 26 August 2020, https://www.darpa.mil/news-events/2020-08-26.

DARPA, 'DARPA initiates design of LongShot Unmanned Air Vehicle,' 8 February 2021, https://www.darpa.mil/news-events/2021–02–08.

Dastin, Jeffrey. 'Amazon scraps secret AI recruiting tool that showed bias against women', *Reuters*, 10 October 2018, https://www.reuters.com/article/us-amazon-com-jobs-automation-insight/amazon-scraps-secret-ai-recruiting-tool-that-showed-bias-against-women-idUSKCN1MK08G.

Davies, Joshua. 'Say hello to Stanley,' *WIRED*, 1 June 2006, https://www.wired.com/2006/01/stanley/.

Dehaene, Stanislas. *Reading in the Brain: The New Science of How We Read*. New York: Penguin Books, 2010.

Dressel, Julia, and Hany Farid. 'The accuracy, fairness, and limits of predicting recidivism', *Science Advances* 4, no. 1 (2018): eaao5580.

Dunbar, Robin I. M. 'Neocortex size and group size in primates: a test of the hypothesis', *Journal of Human Evolution* 28, no. 3 (1995): 287–296.

Eady, Yarrow. 'Tesla's deep learning at scale: using billions of miles to train neural networks', *Towards Data Science*, 7 May 2019, https://towardsdatascience.com/teslas-deep-learning-at-scale-7eed85b235d3.

Eagleman, David. 'Can we create new senses for humans', *TED talks*, March 2015, https://www.ted.com/talks/david_eagleman_can_we_create_new_senses_for_humans?utm_campaign=tedspread&utm_medium=referral&utm_source=tedcomshare.

Elgammal, Ahmed, Bingchen Liu, Mohamed Elhoseiny, and Marian Mazzone. 'Can: Creative adversarial networks, generating "art" by learning about styles and deviating from style norms', *arXiv preprint arXiv:1706.07068* (2017).

Ellsberg, Daniel. *Secrets: A Memoir of Vietnam and the Pentagon Papers*. London: Penguin Books, 2003.

Ellsberg, Daniel. *The Doomsday Machine: Confessions of a Nuclear War Planner*. London, 2018.

Ernest, Nicholas, David Carroll, Corey Schumacher, Matthew Clark, Kelly Cohen, and Gene Lee. 'Genetic fuzzy based artificial intelligence for unmanned combat aerial vehicle control in simulated air combat missions', *Journal of Defense Management* 6, no. 1 (2016): 2167–0374.

Felter, Joseph H., and Jacob N. Shapiro. 'Limiting Civilian Casualties as

Part of a Winning Strategy: The Case of Courageous Restraint', *Daedalus* 146, no. 1 (2017): 44–58.

Fisher, Lucy. 'Stress of killing from afar creates shortage of MoD drone operators', *The Times*, 13 January 2020, https://www.thetimes.co.uk/article/stress-of-killing-from-afar-creates-shortage-of-mod-drone-operators-0fnm36r6t.

Ford, Martin. *Architects of Intelligence: The truth about AI from the people building it*. Birmingham: Packt Publishing Ltd, 2018.

FR24 News, 'France gives [ethical approval] to "augmented soldiers"', 12 December 2020, https://www.fr24news.com/a/2020/12/france-gives-a-sign-of-ethics-to-augmented-soldiers.html.

Frankl, Viktor E. *Man's Search for Meaning: The Classic Tribute to Hope from the Holocaust*. London: Rider, 2004.

Franklin-Wallace, Oliver. 'If we are living in a simulation, this UK start up probably built it,' *WIRED*, 11 May 2017, https://www.wired.co.uk/article/improbable-quest-to-build-the-matrix.

Freedman, Lawrence, and Michaels, Jeffrey. *The Evolution of Nuclear Strategy*. Fourth ed. London: Palgrave Macmillan, 2019.

Fu, Rui, Dawei Yu, Jilong Ren, Chongyang Li, Jing Wang, Guihai Feng, Xuepeng Wang et al. 'Domesticated cynomolgus monkey embryonic stem cells allow the generation of neonatal interspecies chimeric pigs', *Protein & Cell* 11, no. 2 (2020): 97–107.

Fursenko, A. A., and Timothy J. Naftali. *One Hell of a Gamble: Khrushchev, Castro, Kennedy, and the Cuban Missile Crisis, 1958–1964*. London: Pimlico, 1999.

Future of Life Institute, 'Slaughterbots', 13 November 2017, https://youtu.be/HipTO_7mUOw.

Garfinkel, Ben, and Allan Dafoe. 'How does the offense-defense balance scale?' *Journal of Strategic Studies* 42, no. 6 (2019): 736–763.

Geertz, Clifford, and Robert Darnton. *The Interpretation of Cultures: Selected Essays*. New York: Basic Books, 2017.

Gertner, Jon. *The Idea Factory: Bell Labs and the Great Age of American Innovation*. New York: Penguin Press, 2012.

Gladwell, Malcolm. *Outliers: The Story of Success*. London: Allen Lane, 2008.

Gleick, James. *The Information: A History, a Theory, a Flood*. London: Fourth Estate, 2011.

Gottschall, Jonathan. *The Storytelling Animal: How Stories Make Us Human*. Boston Mass.: Houghton Mifflin Harcourt, 2012.

Greene, Joshua D. *Moral Tribes: Emotion, Reason, and the Gap between Us and Them*. New York: Penguin Press, 2013.

Gregory, Andy. 'Coronavirus: Former MI6 head claims pandemic "started as accident" in Chinese laboratory', *The Independent*, 4 June 2020, https://www.independent.co.uk/news/world/asia/coronavirus-chinese-laboratory-wuhan-mi6-richard-dearlove-conspiracy-theories-a9547851.html.

Harkins, Gina. 'Top Marine General: We need to get comfortable with throwaway equipment,' *Military News*, 2 February 2021, https://www.military.com/daily-news/2021/02/02/top-marine-general-we-need-get-comfortable-throwaway-equipment.html.

Heaven, Douglas. 'No limit: AI poker bot is first to beat professionals at multiplayer game', *Nature* 571, no. 7765 (2019): 307–309.

Hennessy, Peter, and James Jinks. *The Silent Deep: The Royal Navy Submarine Service since 1945*. London: Penguin Books, 2016.

Hern, Alex. 'Fitness tracking app Strava gives away location of secret US army bases', *The Guardian*, 28 January 2018, https://www.theguardian.com/world/2018/jan/28/fitness-tracking-app-gives-away-location-of-secret-us-army-bases.

Hillary, Richard. *The Last Enemy*. London: Macmillan, 1942.

Hodges, Andrew. *Alan Turing: The Enigma*. London: Burnett, 1983.

Hoffman, David E. *Dead Hand: The Untold Story of the Cold War Arms Race and Its Dangerous Legacy*. New York: Doubleday, 2009.

Hoffman, F. G., 'Will war's nature change in the seventh military revolution,' *Parameters*, 47:4, 2017/8, pp. 19–31.

Ilyas, Andrew, Logan Engstrom, Anish Athalye, and Jessy Lin. 'Black-box adversarial attacks with limited queries and information', *arXiv preprint arXiv:1804.08598* (2018).

Improbotics: An Improvised Theatre Experiment, https://improbotics.org.

Inoue, Sana, and Tetsuro Matsuzawa. 'Working memory of numerals in chimpanzees', *Current Biology* 17, no. 23 (2007): R1004-R1005.

Insinna, Valerie, 'The US Air Force has built and flown a mysterious full-scale prototype of its future fighter jet,' *Defence News*, 15 September 2020, https://www.defensenews.com/breaking-news/2020/09/15/the-us-air-force-has-built-and-flown-a-mysterious-full-scale-prototype-of-its-future-fighter-jet/.

Jaderberg, Max, Wojciech M. Czarnecki, Iain Dunning, Luke Marris, Guy Lever, Antonio Garcia Castaneda, Charles Beattie et al. 'Human-level performance in 3D multiplayer games with population-based reinforcement learning', *Science* 364, no. 6443 (2019): 859–865.

Jensen, Benjamin and John Paschkewitz, 'Mosaic warfare: small and scalable are beautiful', *War on the Rocks*, 23 December 2019, https://warontherocks.com/2019/12/mosaic-warfare-small-and-scalable-are-beautiful/.

Jiang, Linxing, Andrea Stocco, Darby M. Losey, Justin A. Abernethy, Chantel S. Prat, and Rajesh P. N. Rao. 'BrainNet: a multi-person brain-to-brain interface for direct collaboration between brains', *Scientific Reports* 9, no. 1 (2019): 1–11.

Johnson, James. *The AI Revolution & the Future of Warfare: Great Power Competition & Strategic Stability*. Manchester: Manchester University Press, forthcoming 2021.

Johnson, Lyndon B. *The Vantage Point: Perspectives of the Presidency 1963–1969*. New York: Holt, Reinhart and Winston, 1971.

Kania, Elsa B. 'The pursuit of AI is more than an arms race', *Defense One*, 19 April 2018, https://www.defenseone.com/ideas/2018/04/pursuit-ai-more-arms-race/147579/.

Kania, Elsa B. 'China's ambitions in artificial intelligence: A challenge to the future of democracy?' *Power 3.0*, 8 August 2018, https://www.power3point0.org/2018/08/08/chinas-ambitions-in-artificial-intelligence-a-challenge-to-the-future-of-democracy/.

Kaplan, Fred M. *The Wizards of Armageddon*. Stanford, Calif.: Stanford University Press, 1991.

Kaplan, Fred. *The Bomb: Presidents, Generals, and the Secret History of the Nuclear War* New York: Simon & Schuster, 2020.

Kartchner, Kerry, et al. (ed). *The Routledge Handbook of Strategic Culture*. Abingdon: Routledge, 2021/2 (forthcoming).

Kasparov, Garry K., and Mig Greengard. *Deep Thinking: Where Machine Intelligence Ends and Human Creativity Begins*. London: John Murray, 2018.

Keller, Jared. 'US special operations forces are testing a "guaranteed hit" smart rifle system in Syria', *Task and Purpose*, 5 June 2020, https://taskandpurpose.com/military-tech/special-operations-smart-shooter-syria.

Keller, Laurent, and Elisabeth Gordon. *The Lives of Ants*. Oxford: Oxford UP, 2010.

Kelly, Mary Louise. 'One key, many notes: Ólafur Arnalds' piano rig fuses technology and musicality', *NPR All Things Considered*, 19 July 2018, https://www.npr.org/2018/07/19/630111211/one-key-many-notes-lafur-arnalds-piano-rig-fuses-technology-and-musicality?t=1595412287661.

Kim, H. Jin, Michael I. Jordan, Shankar Sastry, and Andrew Y. Ng. 'Autonomous helicopter flight via reinforcement learning', In *Advances in neural information processing systems*, pp. 799–806. 2004.

King, Anthony. *The Combat Soldier: Infantry Tactics and Cohesion in the Twentieth and Twenty-first Centuries*. Oxford: Oxford University Press, 2013.

Knapton, Sarah. 'Robot that can feel pain invented by scientist', *The Telegraph*, 22 February 2020, https://www.telegraph.co.uk/science/2020/02/22/watch-robot-can-feel-pain-invented-scientists/.

Krulak, Charles, C. 'The Strategic Corporal: Leadership in the three-block war'. *Marines Magazine* 6, 1999, https://apps.dtic.mil/dtic/tr/fulltext/u2/a399413.pdf.

Kumar, Vijay. 'Robots that fly and cooperate', TED Talk, 1 March 2012, https://youtu.be/4ErEBkj_3PY.

Lappin, Yaakov. 'IAI announces new Mini Harpy loitering munition', *Janes*, 19 February 2019, https://www.janes.com/defence-news/news-detail/iai-announces-new-mini-harpy-loitering-munition.

Lappin, Yaakov. 'IDF using Uvision mini loitering munition', *Janes*, 16 July 2020, https://www.janes.com/defence-news/news-detail/idf-using-uvision-mini-loitering-munition.

Le, Quoc V. 'Building high-level features using large scale unsupervised learning', In *2013 IEEE international conference on acoustics, speech and signal processing*, pp. 8595–8598. IEEE, 2013.

Ledford, Heidi. 'CRISPR treatment inserted directly into the body for first time', *Nature* 579, no. 7798 (2020): 185–185.

LeDoux, Joseph E. *Anxious: The Modern Mind in the Age of Anxiety*. London: Oneworld, 2015.

Lei Shi, Xin Luo, Jin Jiang, Yongchang Chen, Cirong Liu, Ting Hu, Min Li, Qiang Lin, Yanjiao Li, Jun Huang, Hong Wang, Yuyu Niu, Yundi Shi, Martin Styner, Jianhong Wang, Yi Lu, Xuejin Sun, Hualin Yu, Weizhi Ji, Bing Su, 'Transgenic rhesus monkeys carrying the human *MCPH1* gene copies show human-like neoteny of brain development', *National Science Review* 6, no. 3 (May 2019): 480–493.

Lenat, Douglas B. 'CYC: A large-scale investment in knowledge infrastructure', *Communications of the ACM* 38, no. 11 (1995): 33–38.

Lepore, Jill. *If Then: How One Data Company Invented the Future*. London: John Murray, 2020.

Leuner, John. 'A Replication Study: Machine Learning Models are capable of predicting sexual orientation from facial images', *arXiv preprint arXiv:1902.10739* (2019).

Lieberman, Matthew D. *Social: Why Our Brains Are Wired to Connect*. Oxford: Oxford University Press, 2013.

Lin, Dayu, Maureen P. Boyle, Piotr Dollar, Hyosang Lee, E. S. Lein, Pietro Perona, and David J. Anderson. 'Functional identification of an aggression locus in the mouse hypothalamus', *Nature* 470, no. 7333 (2011): 221–226.

Lin, Z., J. Jung, S. Goel, and J. Skeem. (2020). 'The limits of human predictions of recidivism', *Science Advances*, 6(7). https://doi.org/10.1126/sciadv.aaz0652.

Lohr, Steve. 'How do you vote? 50 million Google images give a clue', *The New York Times*, 31 December 2017, https://www.nytimes.com/2017/12/31/technology/google-images-voters.html.

Lye, Harry. 'RAF to launch swarming drone squadron in April', *Airforce Technology*, 13 January 2020, https://www.airforce-technology.com/news/raf-swarming-drones/.

Lye, Harry. 'UK flies 20-drone swarm in major test,' *Airforce Technology*, 28 January 2021, https://www.airforce-technology.com/news/uk-flies-20-drone-swarm-in-major-test/.

Maas, Matthijs M. 'How Viable Is International Arms Control for Military Artificial Intelligence? Three Lessons from Nuclear Weapons', *Contemporary Security Policy* 40, no. 3 (2019): 285–311.

Madrigal, Alexis C. 'The computer that predicted the US would win the Cold War', *The Atlantic*, 5 October 2017, https://www.theatlantic.com/technology/archive/2017/10/the-computer-that-predicted-the-us-would-win-the-vietnam-war/542046/.

Marcus, Gary, and Ernest Davis. *Rebooting AI: Building artificial intelligence we can trust*. New York: Pantheon, 2019.

Markwica, Robin. *Emotional Choices: How the Logic of Affect Shapes Coercive Diplomacy*. Oxford: Oxford University Press, 2018.

Marshall, Alex. 'A robot walks into a bar. But can it do comedy?' *The New York Times*, 8 August 2018, https://www.nytimes.com/2018/08/08/arts/ai-comedy-artificial-intelligence-piotr-mirowski.html.

Martin, Christopher, Rahul Bhui, Peter Bossaerts, et al. 'Chimpanzee choice rates in competitive games match equilibrium game theory predictions', *Scientific Reports* 4, 5182 (2015).

Martin, David. 'Russian hacking proves lethal after Ukranian military app hijacked', *CBS News*, 22 December 2016, https://www.cbsnews.com/news/russian-hacking-proves-lethal-after-ukrainian-military-app-compromised/.

Maslow, Abraham. 'A theory of human motivation', *Psychological Review* 50, no. 4 (1943): 370–396.

McCulloch, Warren S., and Walter Pitts. 'A logical calculus of the ideas immanent in nervous activity', *The Bulletin of Mathematical Biophysics* 5, no. 4 (1943): 115–133.

McFarland, Matt. 'Elon Musk: "With Artificial Intelligence we are summoning the demon."', *Washington Post*, 24 October 2014.

Metz, Cade. 'Inside Libratus, the poker AI that out-bluffed the best humans', *WIRED*, 1 February, 2017, https://www.wired.com/2017/02/libratus/.

Metz, Cade. 'The Robots aren't as human as they seem', *The New York Times*, 28 September 2018, https://www.nytimes.com/2018/09/28/technology/robots-human.html.

Michel, Arthur Holland. *Eyes in the Sky: The Secret Rise of Gorgon Stare and How it Will Watch Us All*. New York: Houghton Mifflin Harcourt, 2019.

Mill, John Stuart, and Gertrude Himmelfarb. *On Liberty*. Harmondsworth: Penguin, 1985.

Ministry of Defence, Land Warfare Development Centre, 'Land Operations,' Army Doctrine Publication AC 71940, 31 March 2017, https://assets.publishing.service.gov.uk/government/uploads/system/uploads/attachment_data/file/605298/Army_Field_Manual__AFM__A5_Master_ADP_Interactive_Gov_Web.pdf.

Ministry of Defence, 'JDP 0–30.2: Unmanned Aircraft Systems', Development, Doctrine and Concepts Centre, August 2017, https://assets.publishing.service.gov.uk/government/uploads/system/uploads/attachment_data/file/673940/doctrine_uk_uas_jdp_0_30_2.pdf.

Ministry of Defence, 'Human Machine Teaming', Joint Concept Note 1/18, Development, Concept and Doctrine Centre, May 2018, https://assets.publishing.service.gov.uk/government/uploads/system/uploads/attachment_data/file/709359/20180517-concepts_uk_human_machine_teaming_jcn_1_18.pdf.

Minsky, Marvin, and Seymour Papert. *Perceptrons: An Introduction to Computational Geometry*. Cambridge, Mass: MIT, 1969.

Minsky, Marvin. 'A celebration of Isaac Asimov', *The New York Times*, 12 April 1992, https://www.nytimes.com/1992/04/12/business/technology-a-celebration-of-isaac-asimov.html?pagewanted=all&src=pm.

Mitchell, Kevin J. *Innate: How the Wiring of Our Brains Shapes Who We Are*. Princeton, NJ: Princeton University Press, 2018.

Mizokami, Kyle. 'Become a super sniper: DARPA is turning 0.50 caliber bullets into guided rounds', *The National Interest*, 28 July 2018, https://nationalinterest.org/blog/buzz/become-super-sniper-darpa-turning-50-caliber-bullets-guided-rounds-27101.

Mizokami, Kyle. 'This robot tank is a glimpse at the future of armoured warfare', *Popular Mechanics*, 23 June 2020, https://www.popularmechanics.com/military/weapons/a32947926/type-x-robot-combat-vehicle/.

Mnih, Volodymyr, Koray Kavukcuoglu, David Silver, Andrei A. Rusu, Joel Veness, Marc G. Bellemare, Alex Graves et al. 'Human-level control through deep reinforcement learning', *Nature* 518, no. 7540 (2015): 529–533.

Morstatter, Fred, Aram Galstyan, Gleb Satyukov, Daniel Benjamin, Andres Abeliuk, Mehrnoosh Mirtaheri, K. S. M. Hossain et al. 'SAGE: a hybrid geopolitical event forecasting system', In *Proceedings of the 28th International Joint Conference on Artificial Intelligence*, pp. 6557–6559. AAAI Press, 2019.

Musk, Elon, et al. 'An open letter to the United Nations Convention on Certain Conventional Weapons', 21 August 2017, https://futureoflife.org/autonomous-weapons-open-letter-2017/.

Nagel, Thomas. 'What Is It Like to Be a Bat?' *The Philosophical Review* 83, no. 4 (1974): 435–50.

NASA, 'Borg computer collective designs NASA space antenna', 16 February 2006, https://www.nasa.gov/centers/ames/research/exploringtheuniverse/borg.html.

Neubauer, Catherine, Gregory Gremillion, Brandon S. Perelman, Claire La Fleur, Jason S. Metcalfe, and Kristin E. Schaefer. 'Analysis of facial expressions explain affective state and trust-based decisions during interaction with autonomy.' In *International Conference on Intelligent Human Systems Integration*, pp. 999–1006.

Nilsson, Nils J. *The Quest for Artificial Intelligence: A History of Ideas and Achievements*. Cambridge: Cambridge University Press, 2010.

New Scientist, 'Cyborg insects with wings controlled by humans', March 2008, https://youtu.be/dSCLBG9KeX4.

Oh, Junhyuk, Matteo Hessel, Wojciech M. Czarnecki, Zhongwen Xu, Hado van Hasselt, Satinder Singh, and David Silver. 'Discovering Reinforcement Learning Algorithms', *arXiv preprint arXiv:2007.08794* (2020).

O'Mara, Margaret, *The Code: Silicon Valley and the Remaking of America*. New York: Penguin Press, 2019.

OpenAI, 'OpenAI Five defeats Dota 2 World Champions', 15 April 2019, https://openai.com/blog/openai-five-defeats-dota-2-world-champions/.

Ortiz-Catalan, Max, Enzo Mastinu, Paolo Sassu, Oskar Aszmann, Rickard Brånemark. 'Self-Contained Neuromusculoskeletal Arm Prostheses', *New England Journal of Medicine* 382, no. 18 (2020): 1732.

Osborn, Kris, 'Future of war will be "hyperactive battlefields": US Army General,' *The National Interest*, 30 January 2021, https://nationalinterest.org/blog/buzz/future-war-will-be-'hyperactive-battlefields'-us-army-general-177371.

Osinga, Frans. *Science, Strategy and War: The Strategic Theory of John Boyd*. Abingdon: Routledge, 2007.

Overy, Richard J. *Battle of Britain: The Myth and Reality*. New York: Norton (W.W) & Co, 2002.

Overy, Richard J. *The Bombing War: Europe 1939–1945*. London: Allen Lane, 2014.

Pais-Vieira, Miguel, Mikhail Lebedev, Carolina Kunicki, Jing Wang, and Miguel A. L. Nicolelis. 'A brain-to-brain interface for real-time sharing of sensorimotor information', *Scientific reports* 3 (2013): 1319.

Paret, Peter, Gordon Alexander Craig, and Felix Gilbert. *Makers of Modern Strategy: From Machiavelli to the Nuclear Age*. Princeton, N.J.: Princeton University Press, 1986.

Pawlyk, Oriana. 'The Air Force risks a drone pilot shortage, GAO finds', *Military.com*, 26 June 2020, https://www.military.com/daily-news/2020/06/26/air-force-risks-drone-pilot-shortage-gao-finds.html.

Payne, Kenneth. *The Psychology of Modern Conflict: Evolutionary Theory, Human Nature and a Liberal Approach to War*. Basingstoke: Palgrave Macmillan, 2015.

Payne, Kenneth. *Strategy, Evolution, and War: From Apes to Artificial Intelligence*. Washington, DC: Georgetown University Press, 2018.

Payne, Kenneth. 'Artificial intelligence: a revolution in strategic affairs?' *Survival* 60, no. 5 (2018): 7–32.

Perrow, Charles. *Normal Accidents: Living with High-risk Technologies*. New York: Basic, 1984.

Plomin, Robert. *Blueprint: How DNA Makes Us Who We Are*. London: Penguin, 2019.

Poldrack, Russell A. *The New Mind Readers: What Neuroimaging Can and Cannot Reveal about Our Thoughts*. Princeton, N.J.: Princeton University Press, 2018.

Poundstone, William. *Prisoner's Dilemma: John von Neumann, Game Theory, and the Puzzle of the Bomb*. New York: Anchor Books, 1992.

President's Foreign Intelligence Advisory Board, 'The Soviet "War Scare"' 15 February 1990. Declassified October 2015. https://nsarchive2.gwu.edu/nukevault/ebb533-The-Able-Archer-War-Scare-Declassified-PFIAB-Report-Released/2012-0238-MR.pdf.

Rawls, John. *A Theory of Justice*. Boston, MA: Harvard University Press, 2009.

Regalado, Antonio. 'Exclusive: Chinese scientists are creating CRISPR babies', *MIT Technology Review*, 25 November 2018, https://www.technologyreview.com/2018/11/25/138962/exclusive-chinese-scientists-are-creating-crispr-babies/.

Rid, Thomas. *Cyber War Will Not Take Place*. New York: Oxford University Press, USA, 2013.

Rid, Thomas. *Rise of the Machines: The Lost History of Cybernetics*. New York: WW Norton, 2016.

Robinson, Heath. *Heath Robinson's Second World War*. Oxford: The Bodleian Library, 2016.

Roland, Alex, and Philip Shiman. *Strategic Computing: DARPA and the Quest for Machine Intelligence, 1983–1993*. Cambridge, Mass: MIT Press, 2002.

Russell, Stuart J. *Human Compatible: Artificial Intelligence and the Problem of Control*. London: Allen Lane, 2019.

Russon, Mary-Ann. 'The robot that watches as you cross the road', *BBC News*, 28 August 2018, https://www.bbc.co.uk/news/business-45330451.

Sagar, Mark, Mike Seymour, and Annette Henderson. (2016). 'Creating connection with autonomous facial animation', *Communications of the ACM*. 59. 82–91. 10.1145/2950041.

Salisbury, Emma. 'A cautionary tale on ambitious feats of AI: The Strategic Computing program', *War on the Rocks*, 22 May 2020, https://warontherocks.com/2020/05/cautionary-tale-on-ambitious-feats-of-ai-the-strategic-computing-program/.

Sandlin, Destin. 'The backward bicycle brain', *Smarter Every Day*, 24 April 2015, https://youtu.be/MFzDaBzBlL0.

Sanger, David E. *The Perfect Weapon*. London: Scribe, 2018.

Sato, Hirotaka, Christopher W. Berry, Brendan E. Casey, Gabriel Lavella,

Ying Yao, John M. VandenBrooks, and Michel M. Maharbiz. 'A cyborg beetle: insect flight control through an implantable, tetherless microsystem', In *2008 IEEE 21st International Conference on Micro Electro Mechanical Systems*, pp. 164–167. IEEE, 2008.

Scharre, Paul. 'How swarming will change warfare', *Bulletin of the Atomic Scientists* 74, no. 6 (2018): 385–389.

Scharre, Paul, 'Autonomous weapons and stability', unpublished PhD thesis, King's College London, 2020.

Schelling, Thomas C. *The Strategy of Conflict*. New York: Oxford University Press, 1963.

Schelling, Thomas C. *Arms and Influence*. 2008. New Haven: Yale University Press, 1966.

Schrittwieser, Julian, Ioannis Antonoglou, Thomas Hubert, Karen Simonyan, Laurent Sifre, Simon Schmitt, Arthur Guez et al. 'Mastering Atari, go, chess and shogi by planning with a learned model', *arXiv preprint arXiv:1911.08265* (2019).

Seabrook, John. 'The next word: Where will predictive text take us?' *The New Yorker*, 14 October 2019 https://www.newyorker.com/magazine/2019/10/14/can-a-machine-learn-to-write-for-the-new-yorker.

Selfridge, Oliver G. 'Pandemonium: a paradigm for learning', *Mechanism of Thought Processes: Proceedings of a Symposium Held at the National Physical Laboratory* (1958): 513–526.

Senator, Ted. 'Serial Interactions in Imperfect Information Games Applied to Complex Military Decision Making (SI3-CMD)', DARPA program information, undated, https://www.darpa.mil/program/serial-interactions-in-imperfect-information-games-applied-to-complex-military-decision-making.

Silver, David, Aja Huang, Chris J. Maddison, Arthur Guez, Laurent Sifre, George Van Den Driessche, Julian Schrittwieser et al. 'Mastering the game of Go with deep neural networks and tree search', *Nature* 529, no. 7587 (2016): 484–489.

Simon, Herbert A. *The Sciences of the Artificial*. 3rd edn London: MIT Press, 1996.

Singer, Natasha. 'Amazon is pushing facial technology that a study says could be biased', *The New York Times*, 24 January 2019, https://www.

nytimes.com/2019/01/24/technology/amazon-facial-technology-study.html.

Sini, Rozina. 'Does Saudi robot citizen have more rights than women?' *BBC News*, 26 October 2017, https://www.bbc.co.uk/news/blogs-trending-41761856.

Slovic, Paul, Melissa L. Finucane, Ellen Peters, and Donald G. MacGregor. 'Risk as analysis and risk as feelings: Some thoughts about affect, reason, risk, and rationality', *Risk Analysis: An International Journal* 24, no. 2 (2004): 311–322.

Smith, Adam. 'Cognitive empathy and emotional empathy in human behavior and evolution', *The Psychological Record* 56, no. 1 (2006): 3–21.

Sofge, Eric. 'The DARPA Robotics Challenge was a bust', *Popular Science*, 6 July 2015, https://www.popsci.com/darpa-robotics-challenge-was-bust-why-darpa-needs-try-again/.

Sorensen, Theodore C. *Decision-making in the White House: The Olive Branch or the Arrows*. New York: Columbia University Press, 1963.

Sorensen, Theodore C. *Kennedy*. New York: Harper Perennial, 2009.

Stahl, Jennifer. 'Could Google be the world's next great choreographer?' *Dance Magazine*, 11 January 2019, https://www.dancemagazine.com/is-google-the-worlds-next-great-choreographer-2625652667.html.

Stewart, Jack. 'Israel's self-flying "Cormorant" whisks soldiers to safety', *WIRED*, 26 May 2018, https://www.wired.com/story/tactical-robotics-cormorant-autonomous/.

Stockton, Nick. 'Woman controls a fighter jet sim using only her mind', *WIRED*, 5 March 2015, https://www.wired.com/2015/03/woman-controls-fighter-jet-sim-using-mind/.

Surdu, John R., and Kevin Kittka. 'Deep Green: Commander's tool for COA's Concept', In *Proceedings of Computing, Communications and Control Technology Conference*, vol. 29. 2008.

Strassler, Robert B. , and Richard Crawley, *The Landmark Thucydides: A Comprehensive Guide to the Peloponnesian War*. London: Free Press, 1996.

Tass Russian News Agency, 'Putin stresses whoever takes the lead in artificial intelligence will rule world', 1 September 2017, https://tass.com/society/963209.

Tetlock, Philip E. *Expert Political Judgment: How Good Is It? How Can We Know?* Princeton, NJ: Princeton UP, 2005.

Tetlock, Philip E., and Dan Gardner. *Superforecasting: The Art and Science of Prediction*. London: Random House, 2015.

Tetlock, Philip E., Barbara A. Mellers, and J. Peter Scoblic. 'Bringing probability judgments into policy debates via forecasting tournaments', *Science* 355, no. 6324 (2017): 481–483.

Thrun, Sebastian, Mike Montemerlo, Hendrik Dahlkamp, David Stavens, Andrei Aron, James Diebel, Philip Fong et al. 'Stanley: The robot that won the DARPA Grand Challenge', *Journal of Field Robotics* 23, no. 9 (2006): 661–692.

Thys, Simen, Wiebe Van Ranst, and Toon Goedemé. 'Fooling automated surveillance cameras: adversarial patches to attack person detection', In *Proceedings of the IEEE Conference on Computer Vision and Pattern Recognition Workshops*, pp. 49–55. 2019.

Trivers, Robert L. 'The evolution of reciprocal altruism', *The Quarterly Review of Biology* 46, no. 1 (1971): 35–57.

Tucker, Patrick. 'Marines are building robotic war balls', *Defense One*, 12 February 2015, https://www.defenseone.com/technology/2015/02/marines-are-building-robotic-war-balls/105258/.

Turing, Alan Mathison. 'On computable numbers, with an application to the Entscheidungsproblem.' *J. of Math* 58, no. 345–363 (1936): 5.

Turing, Alan Mathison. 'Computing machinery and intelligence', *Mind*, Volume LIX, Issue 236, October 1950, pp. 433–460, https://doi.org/10.1093/mind/LIX.236.433.

Twilley, Nicola. 'Seeing with your tongue,' *The New York Times*, 8 May 2017, https://www.newyorker.com/magazine/2017/05/15/seeing-with-your-tongue.

United States National Security Commission on Artificial Intelligence, 'Draft Final Report,' January 2021, https://www.nscai.gov/wp-content/uploads/2021/01/NSCAI-Draft-Final-Report-1.19.21.pdf.

United States Navy, 'Perdix Swarm demo', 12 Jan 2017, https://youtu.be/bsKbGc9TUHc.

Verbruggen, Maaike, 'The question of swarms control: challenges to ensuring human control over military swarms', SIPRI Non Proliferation and Disarmament Papers, no. 65, December 2019, https://www.sipri.org/sites/default/files/2019–12/eunpdc_no_65_031219.pdf.

Vigen, Tyler. *Spurious Correlations*. New York: Hachette Books, 2015.

Walzer, Michael. *Just and Unjust Wars: A Moral Argument with Historical Illustrations*. New York: Basic Books, 1977.

Wang, Yilun, and Michal Kosinski. 'Deep neural networks are more accurate than humans at detecting sexual orientation from facial images', *Journal of Personality and Social Psychology* 114, no. 2 (2018): 246.

Westwick, Peter. *Stealth: The Secret Contest to Invent Invisible Aircraft*. Oxford: Oxford University Press, 2019.

Whetham, David, and David Fisher. *Ethics, Law and Military Operations*. Basingstoke: Palgrave Macmillan, 2011.

White, Ralph K. *Fearful Warriors: A Psychological Profile of U.S.-Soviet Relations*. London: Collier Macmillan, 1984.

Wiener, Norbert. *Cybernetics: Or, Control and Communication in the Animal and the Machine*. Second edn, Mansfield Centre, CT: Martino Publishing, 2013.

Wittgenstein, Ludwig, translated by D.F. Pears and B.F. McGuinness, *Tractatus Logico-philosophicus*. London: Routledge, 2003.

Wohlstetter, Albert. 'The delicate balance of terror', *Foreign Affairs* 37 (1958): 211.

Wong, Julia Carrie. 'The FBI and Apple are facing off over an iPhone again. What's going on?' *The Guardian*, 15 January 2020, https://www.theguardian.com/us-news/2020/jan/14/fbi-apple-faceoff-iphone-florida-shooting.

Wong, Yuna Huh, John Yurchak, Robert W. Button, Aaron Frank, Burgess Laird, Osonde A. Osoba, Randall Steeb, Benjamin N. Harris, and Sebastian Joon Bae, *Deterrence in the Age of Thinking Machines*. Santa Monica, CA: RAND Corporation, 2020. https://www.rand.org/pubs/research_reports/RR2797.html.

Work, Robert O. 'Establishment of an Algorithmic Warfare Cross-Functional Team (ProjectMaven)', US Department of Defense Memorandum, 26 April 2017, https://www.govexec.com/media/gbc/docs/pdfs_edit/establishment_of_the_awcft_project_maven.pdf.

Wrangham, Richard W. *The Goodness Paradox: The Strange Relationship between Virtue and Violence in Human Evolution*. New York: Vintage Books, 2019.

Wright, Robert. *The Moral Animal: Evolutionary Psychology and Everyday Life*. London: Little, Brown, 1995.

Zhang, Shaomin, Sheng Yuan, Lipeng Huang, Xiaoxiang Zheng, Zhaohui Wu, Kedi Xu, and Gang Pan. 'Human mind control of rat cyborg's continuous locomotion with wireless brain-to-brain interface', *Scientific Reports* 9, no. 1 (2019): 1–12.

ACKNOWLEDGEMENTS

For their insights on Artificial Intelligence in war, I'm hugely grateful to a small community of scholars working in the field. In particular, I've benefited from reading and talking with Al Brown, Keith Dear, Theo Farrell, Ulrike Franke, David Galbreath, Michael Horowitz, Jim Johnson, Elsa Kania, Matthijs Maas, Mick Ryan, Emma Salisbury, Paul Scharre and Maaike Verbruggen. Special thanks to my Royal Marine student Chris whose great insights really sharpened my own writing. And thanks also to Frank Hoffman—a brilliant strategic thinker and generous scholar.

At KCL, Huw Davies badgered me into writing an online MA course on AI and strategy that seeded the ideas here. Chris Tripodi and Jon Hill are colleagues and firm friends who continually inspire and energise my research. DeepMind's Kareem Ayoub got me started on all this five years ago, and remains a close friend, even if these days AI is the last thing we talk about. Michael Dwyer, my publisher, has been terrific throughout. This is our second book together and hopefully not our last. Euan Thorneycroft at AM Heath gave me some excellent advice early in the process.

I wrote most of this book while flat on my back recovering from Covid-19. Suddenly AI didn't seem like the biggest threat to humanity. I'm so grateful for my friends and family for keeping me sane and looking after me. Thanks especially to Charli Hunt, Jimmy Hillis, Richard Exley and Claire Milner for some good fun in tough times. The book is dedicated to two great friends, Stephen Hare and Stephanie Jones, in honour of adventures past and adventures to come.

INDEX

INDEX

INDEX

INDEX